Urban Food Security in a Crisis Prone World

Atif A. Kubursi • Nathaniel K. Newlands
Amani Alfarra
Editors

Urban Food Security in a Crisis Prone World

The Urban, Water, and Food Nexus

Springer

Editors
Atif A. Kubursi
McMaster University
Burlington, ON, Canada

Nathaniel K. Newlands
Agriculture and Agri-Food Canada
Summerland, BC, Canada

Amani Alfarra
UN-FAO
Rome, Italy

ISBN 978-3-031-89439-8 ISBN 978-3-031-89440-4 (eBook)
https://doi.org/10.1007/978-3-031-89440-4

This Springer imprint is published by the registered company Springer Nature Switzerland AG
The registered company address is: Gewerbestrasse 11, 6330 Cham, Switzerland

As editors we dedicate this book to the late Professor Donald Houessou in recognition of his many contributions to food security issues

Foreword

Global agrifood systems have experienced many crises in recent years, most notably the COVID-19 pandemic, the conflict in Ukraine, including extreme weather events (e.g., heatwaves, drought, floods). These multiple crises have collectively driven-up prices for food and agricultural inputs, increasing food insecurity, malnutrition, poverty rates, alongside other adverse effects across many countries. The related social and economic impacts from the crises stand to cause devastating setbacks to achieve the 2030 Agenda for Sustainable Development. The global crises revealed that the world's shock-sensitive agrifood systems rest on increasingly fragile foundations. There is a growing consensus on the need for urgent, collective action from governments, international organizations, the private sector, farmers, and other agrifood system stakeholders to build resilience necessary to safeguard food supply chains and protect food access and availability.

This book brings together noted scholars and experts in disparate fields such as economists, engineers, physical scientists, agronomists, social scientists, data scientists, and environmentalists providing their multiple perspectives on the crisis-prone world, and the impacts of these crises on urban food security. It provides analyses of the new global challenges that have resulted from the COVID-19 pandemic and the conflict in Ukraine, combined with climate change and its impacts. It discusses lessons-learned and potential solutions going forward on how to redesign and transform food systems to be more resilient to disruption and disaster. The book also draws on practical case studies from Canada and other countries on several topics including environmental resilience, agricultural best practices, self-provisioning, urban agriculture, and aligning urban farming with green city aspirations. The book explores the potential benefits and challenges of developing strategies to reduce urban risks for producers and support resilience, as an increasing number of agricultural technologies are available to support small-scale producers. This book makes an important contribution to understanding the fragility of urban agriculture to global crises, and the urgent actions and efforts from various stakeholders to bolster food security and resilience of agrifood systems.

The Food and Agriculture Organization of the United Nations (FAO) has been at the forefront in supporting countries to address challenges by developing comprehensive and holistic approaches to proactively and sustainably address the socio-economic impacts arising from global crises while strengthening the long-term resilience of food systems and livelihoods. FAO's Strategic Framework 2022–2031 was developed in the context of major global and regional challenges in the areas of FAO's mandate. Canada and the FAO have worked together for many decades toward finding solutions to many of the most pressing problems of our time, to target long-term development as well as peace, security, and humanitarian assistance. Canada is one of FAO's key resource partners and its assistance has strongly focused on three priority areas: increasing food security, stimulating sustainable economic growth, and securing the future of children and youth, which are well aligned with FAO's Strategic Framework. FAO and Canada stand firmly committed to transforming agrifood systems to make them more efficient, more inclusive, more resilient, and more sustainable for better production, better nutrition, a better environment, and a better life for all, now and for future generations.

Director of Research, Development and Technology Transfer Rachid El Hafid
Summerland and Agassiz Research and Development Centres
Agriculture and Agri-Food Canada (AAFC)
Canada

Director of Land and Water Division Lifeng Li
Food Agriculture Organization of the United Nations (FAO)
Italy

Contents

Contributors

Daryoush Alipour is an Associate Professor at Bu-Ali Sina University in Iran and works on rumen microbiology, using agro-industrial co-products in ruminant nutrition and sustainable ruminant production.

Marta Antonelli is the Director of food4future initiative, Global Footprint Network.

Augustin K. N. Aoudji is the Vice-dean at the Faculty of Agricultural Sciences of the University of Abomey-Calavi, Republic of Benin. He holds a PhD in Agricultural Economics and has extensive research experience on issues related to value chain analysis, food security, agribusiness, and agricultural entrepreneurship.

Sisir Bhandari holds an Erasmus Mundus Global Studies graduate degree and worked as a Junior Research Fellow (from October 2021 to August 2022) at United Nations University—Institute on Comparative Regional Integration Studies (UNU-CRIS). He is currently an International Economic Relations graduate student at American University, where he works as a Research Assistant of Professor Randall Henning on regime complexes in global governance.

Amani Alfarra is a Water Officer in the Land and Water Division of the United Nations Food and Agriculture Organization. She holds a PhD in Water Resources Management from Karlsruhe Institute for Technology, Germany. Her work and research focused on addressing interconnected challenges, especially in the food-water-energy nexus. Specializing in natural resources management, particularly in river basins and transboundary rivers, Dr. Alfarra is dedicated to tackling global issues like climate change. Her expertise lies in water-based strategies for climate change adaptation, emphasizing Nature-based solutions. Her contributions embraced a multidisciplinary, ecosystem-based approach to address water management and integrated watershed issues.

Monika A. Gorzelak is a Research Scientist with Agriculture and Agri-Food Canada in Lethbridge, Alberta, focusing on mycorrhizal ecology and soil bacterial and fungal communities in sustainable agricultural practices.

John R. Groenewegen is the President of JRG Consulting Group, a Guelph, Ontario based firm that provides consulting services to the Canadian agriculture and food sector. John has a doctorate degree in Agriculture and Applied Economics from the University of Minnesota, and a B.Sc. and MA degree from the University of Guelph. Prior to establishing JRG Consulting Group in 1997, John was a Partner with a major multinational consulting firm.

M. Donald Houessou passed away in February 2023; he was a Program director at ACED, a policy-research institution that generates and shares evidence to inform policymaking on food security and nutrition issues in West Africa. He held a PhD in Agricultural and Food Economics from Athena Institute of the VU University of Amsterdam, Netherlands. His research interests extended to food security, community resilience, fisheries, and urban food systems issues.

Nabil Kamal M.Eng., is the President of PerTech Industries Canada Inc, a Burlington Ontario based firm with branches in several countries including Jordan, Saudi Arabia, and the Far East. He is also an inventor with several registered patents and also an international consultant in the areas of food technology, industrial plants, and innovation policy.

Ali Kiani is an Animal Scientist at the Natural Resources Institute Finland (LUKE), in Jokioinen, Finland, specializing in animal biology, animal nutrition, and animal husbandry systems.

Roland Kröbel is a Research Scientist with Agriculture and Agri-Food Canada in Lethbridge, Alberta, working on the whole-farm model, Holos, to permit Canadian farmers to estimate management practice options to reduce their greenhouse gas emissions, and more broadly on the sustainability of Canadian agriculture.

Atif A. Kubursi is an Emeritus Professor of Economics at McMaster University. Dr. Kubursi was Visiting Scholar at University of Cambridge in 1974/75, Member of the Institute for Social and Economic Policy in the Middle East at Harvard University, 1987-1997, and Acting as Executive Secretary of UNESCWA, 2006-2007. He is also President of Econometric Research Ltd., Burlington, Ontario, Canada, and Adjunct Professor at York University School of Environmental and Urban Change.

Karen Landman is a Professor in the School of Environmental Design & Rural Development at the University of Guelph and studies green infrastructure of many types, including urban agriculture.

David Lin is the Chief Science Officer, Global Footprint Network.

Rod J. MacRae retired, was an Associate Professor at York University's Faculty of Environmental and Urban Change in Toronto, Canada. His website, foodpolicyforcanada.info.yorku.ca, details the many strategies required to create a sustainable, health-promoting, and just food systems for Canada. He has a PhD from McGill University.

Dina A. Mandour is an Assistant Professor in the Economics Department at the Faculty of Economics & Political Science, Cairo University, Egypt. She holds a PhD in Economics from the University of Bonn, Germany. Her main fields of interest include computable general equilibrium (CGEs) models, food security, gender, political economy, and health policy. She also worked in several economic policy-oriented centers including the Egyptian Center for Economic Studies and Social Research Center at the American University in Cairo.

Mohammad Reza Marami Milani is a Research Associate at the University of Manitoba and Agriculture and Agri-Food Canada (AAFC) in Lethbridge, Canada, specializing in agricultural data analysis and modelling.

Nidhi Nagabhatla is a Senior Fellow and Cluster Coordinator: Nature, Climate & Health at United Nations University (CRIS), Belgium. With 20 years of work experience, she has led, coordinated, and implemented transdisciplinary projects in various geographical regions of Asia, Africa, Europe, and the Americas. She is affiliated with several universities including Oxford (UK) and Leibniz (Germany), and McMaster University (Canada).

Nathaniel K. Newlands is a Research Scientist with Agriculture and Agri-Food Canada (AAFC) in Summerland, British Columbia, Canada. He is an Adjunct Professor in Geography with the University of Victoria. His research work addresses public-good food-water-energy nexus issues, and tackles broad, integrated, complex global problems to help support and advance global sustainable development. He is a co-chair of the Group on Earth Observations (GEO)'s Disaster Risk Reduction Working Group and President of the International Environmetrics Society (TIES).

Elham Rahmani is an Agricultural Research Data Analyst at Agriculture and Agri-Food Canada (AAFC) in Lethbridge, Canada, with a background in agro-meteorology and big data analysis and modelling.

Karen Schwartzkopf-Genswein is a Principal Research Scientist with Agriculture and Agri-Food Canada in Lethbridge Alberta and with special expertise in physiological and behavioral methods to assess animal stress for use in developing strategies to improve beef cattle welfare.

Robert Smith is the Principal of Midsummer Analytics, an Ottawa-based consultancy. An environmental economist/statistician with more than 30 years' of experience, his work focuses on the measurement of natural capital and its use as a basis for measuring sustainable development. Until 2013, Mr. Smith worked at Statistics Canada, where he led its environment statistics program.

Ben G. J. S. Sonneveld is an Associate Professor at the VU University of Amsterdam and Deputy Director of the Amsterdam Centre for World Food Studies. He holds a PhD in Environmental Economics from the VU University of Amsterdam, Netherlands. His main research interests are in the modelling of the impact of land and water degradation on agricultural production and eco-services. Dr. Sonneveld holds visiting professorships at the Abomey Calavi University, Benin, and at the Al Quds University, Palestine territories.

Emma C. Stephens is a Bio-economist at AAFC-Lethbridge RDC, studying economic factors that influence the adoption of sustainable agricultural practices, as well as the relationships between agricultural science, innovation, and food security outcomes.

Gaurav Thapak is a Researcher and an Urban Planner working with national and local governments and NGOs in India, largely on urbanization and climate action through policy research and field activities. Currently, he works as a Consultant with the State Policy and Planning Commission of Madhya Pradesh, India.

Alban Thomas is a Senior Researcher and Deputy Scientific Director for the environment at INRAE (French Research Institute for Agriculture, Food and the Environment). He holds a PhD in Econometrics; his research topics include the analysis of environmental impacts of agriculture, water management, and transitions toward global food security. He has supervised several policy-oriented reports on these issues for many ministries and international institutions.

Mathis Wackernagel created the *footprint* concept in the early 1990s (with Bill Rees) to compare human demand against planetary or regional ecosystem regeneration. This tool allows researchers to track global overshoots and countries' ecological deficits. In 2003, he founded Global Footprint Network, a sustainability think-tank, making planetary constraints relevant to decision-making.

Leo Wambersie is a Research Associate, Global Footprint Network and is currently a PhD candidate at École de technologie supérieure, Montréal.

Chapter 1
Crisis Prone World Urban Food–Water Nexus: An Introduction

Atif A. Kubursi, Amani Alfarra, and Nathaniel K. Newlands

Abstract Globalization has brought the world tightly together, and computer and information technologies have compressed time and space. The COVID pandemic ravaged economies, disrupted global supply chains, claimed many lives, stressed the over-crowded health systems everywhere, and destroyed numerous livelihoods. At the same time, it uncovered weaknesses and allowed the world a glimpse of a few opportunities. Now more than ever collective and cooperative action is needed to defeat emergent and cascading global crises. We outline ways the global food system can be fundamentally redesigned and restructured to actively tackle climate change and biodiversity loss, promote human health, and reduce food insecurity, improve the resiliency of communities while reducing the overall costs of the system to society. This can be achieved by reconnecting communities with local food production, by changing the way we grow food, design food products, and handle by-products and waste. Through a nexus approach to redesign and restructuring, we can create a less wasteful system by promoting a circular economy for food that provides key benefits for society and the environment by improving nutritional health, reducing the ecological footprint, securing access, mitigating risks of supply defaults, and generating local incomes and jobs that support cities and local communities.

A. A. Kubursi (✉)
Department of Economics Kenneth Taylor Hall (KTH), McMaster University, West Hamilton, ON, Canada
e-mail: kubursi@mcmaster.ca

A. Alfarra
Land and Water (NSL) Division, Food and Agriculture Organization of the United Nations, Rome, Italy

Walter G. Booth School of Engineering Practice and Technology, Hamilton, ON, Canada
e-mail: Amani.Alfarra@fao.org

N. K. Newlands
Agriculture and Agri-Food Canada, Summerland Research and Development Centre, Summerland, BC, Canada

Department of Geography, University of Victoria, Greater Victoria, BC, Canada
e-mail: nathaniel.newlands@agr.gc.ca

A. A. Kubursi et al. (eds.), *Urban Food Security in a Crisis Prone World*,
https://doi.org/10.1007/978-3-031-89440-4_1

Keywords Circular economy · Crises · Global food systems · Nexus approach · Redesign

Introduction

The corona pandemic has ravaged economies, disrupted global supply chains, claimed many lives, stressed the over-crowded health systems everywhere, and destroyed numerous livelihoods. At the same time it uncovered weaknesses and allowed the world a glimpse of a few opportunities to take advantage of them. Its impacts are likely to accelerate in the short and medium terms and there is no sign yet of its end in the near future.

There is an increasing concern that it might continue to impinge on the global economy and on the global and national food systems everywhere. The impact of a massive global economic shock is likely to trickle down to fragile economies and vulnerable communities, exacerbating existing food insecurity, unemployment rates, income and wealth inequalities, and poverty.

Before the pandemic almost 821 million people, representing around 10% of the world's population, were believed to be suffering from hunger, while another 690 million were declared undernourished (8.9%), thus adding up 60 million to these ranks since 2014.[1]

The global food system was already experiencing problems before the onset of the pandemic in 2020. The pandemic created new challenges and difficulties for all aspects of the world food system by affecting availability, accessibility, and affordability. By the end of 2020 and as a result of the ongoing repercussions of the pandemic, food supply shortages were experienced by an additional 260 million people.[2]

Considering the sustained force and isolation dynamics of COVID 19, strengthening the food production and distribution systems are now critical to fighting hunger and tackling the double burden of malnutrition. Policymakers in international organizations and governments everywhere are assessing and responding to its potential negative impacts on people's life and livelihoods, global food trade, domestic food markets, food supply chains and livestock, accessibility, and affordability. Strengthening structurally the accessibility and stability of food systems in the long run is now a priority for many governments and nongovernmental organizations and so is improving affordability and containing food price increases. The United Nations Food and Agriculture Organization (FAO) is at the forefront of this struggle to contain the negative consequences and the damage COVID 19 has wrought on the world and possible corrective measures and programs.

[1] FAO (2020) and GHI (2020).

[2] Ling Ma et al. (2021).

The war in Ukraine coming on the heels of the pandemic exacerbated the pandemic's negative impacts on many levels but particularly on food security and inflationary expectations. The disruption of the supply chains of energy and food are now manifesting their drastic impacts given that Russia and Ukraine are among the largest food producers and exporters. More specifically, Russia is among the three largest oil producers and the second largest oil exporter. Russia produced 11 million barrels of oil daily and exported 4–5 million barrels per day of crude oil and 2–3 million barrels per day of refined oil products.[3] Europe was the main destination of these exports, so much so that on the eve of the war, 25% of Europe's oil imports came from Russia and about 48% of its gas imports.

Energy supplies were supposed to be exempt from sanctions and disruptions but the war raised insurance rates and militated against tankers' normal shipping routes.[4] Moreover, banks were reluctant to issue letters of credit to oil importers to finance their imports from Russia fearing of running afoul of the sanctions. It is now believed that 70% of oil imports to Europe from Russia would be cut.[5]

Energy supply chains were not the only sector that was negatively affected by the geopolitical conflict in Europe; food supply chains were equally affected. Russia and Ukraine accounted for 26.5% of wheat trade, 17% of corn, and about 70% of sunflower (Fig. 1.1) before the war. Russia is the world's largest exporter of wheat, and Ukraine is the fifth largest. Together, they provide 19% of the world's barley supply, 14% of wheat, and 4% of maize, making up more than one-third of global cereal exports. They are also lead suppliers of rapeseed and account for 52% of the world's sunflower oil export market. The global fertilizer supply is also highly concentrated, with Russia as the lead producer.[6]

Availability of these exports was not the only challenge faced by countries dependent on food imports from Russia and Ukraine.[7] Availability was compounded by massive increases in the costs of these products, even when available food supplies are no longer affordable by millions, but particularly in the Global South.[8]

---

[3] https://www.iea.org/reports/russian-supplies-to-global-energy-markets/oil-market-and-russian-supply-2

[4] Project-syndicate.org/commentary/eu-embargo-on-russian-oil-limited-impact-by-daniel-gros-2022-06

[5] https://www.csis.org/analysis/european-union-imposes-partial-ban-russian-oil#:~:text=A1%3A%20On%20June%203%2C%20the,as%20of%20February%205%2C%202023

[6] https://www.fao.org/russian-federation/news/news-detail/New-Scenarios-on-Global-Food-Security-based-on-Russia-Ukraine-Conflict/en

[7] This is especially true for some fifty countries that depend on Russia and Ukraine for 30% or more of their wheat supply. Many of them are least developed countries or low-income, food-deficit countries in Northern Africa, Asia, and the Near East. (UNFAO:Ibid.) (Fig. 1.2)

[8] Food prices already on the rise since the second half of 2020, reached an all-time high in February 2022 due to high demand, input and transportation costs, and port disruptions. Global prices of wheat and barley, for example, rose 31% over the course of 2021 (Fig. 1.3). Rapeseed oil and sunflower oil prices rose more than 60%. High demand and volatile natural gas prices have also driven up fertilizer costs. For instance, the price of urea, a key nitrogen fertilizer, has increased more than threefold in the past 12 months. (UNFAO, Ibid.,)

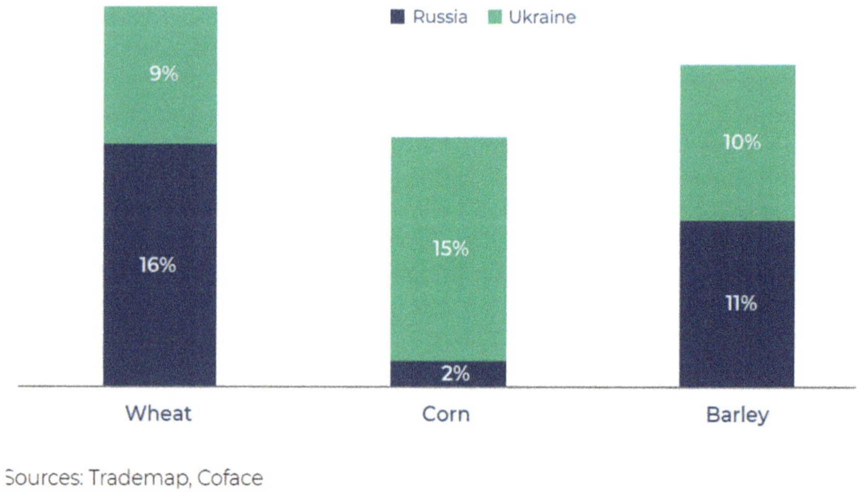

Fig. 1.1 Shares of Russia and Ukraine in food exports

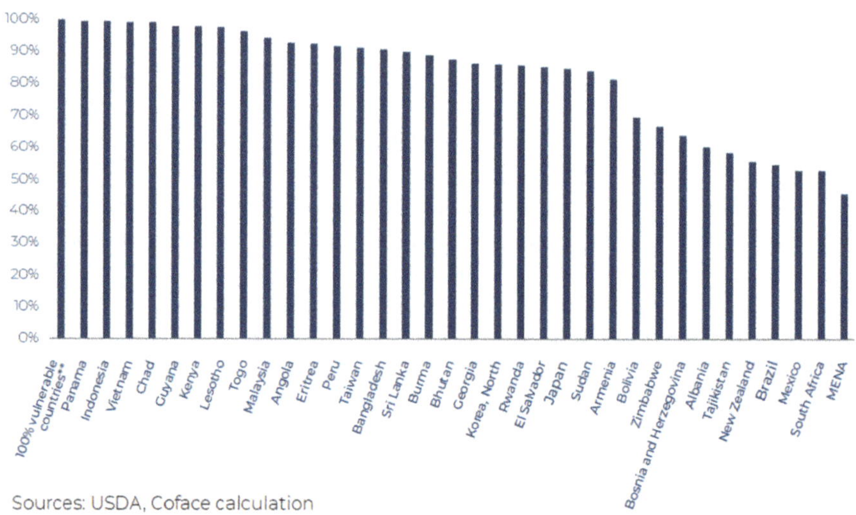

Fig. 1.2 Most vulnerable countries importing wheat as a share of total consumption

Oil prices almost increased sixfold from the beginning of the pandemic, natural gas prices more than quadrupled, and the prices of fertilizers escalated to new heights and are still rising monthly.

The new reality imposed by the pandemic and the war has changed the way people access their food, how food is produced, what food to produce, how costly has it become, and ways and means of rationalizing its production, consumption,

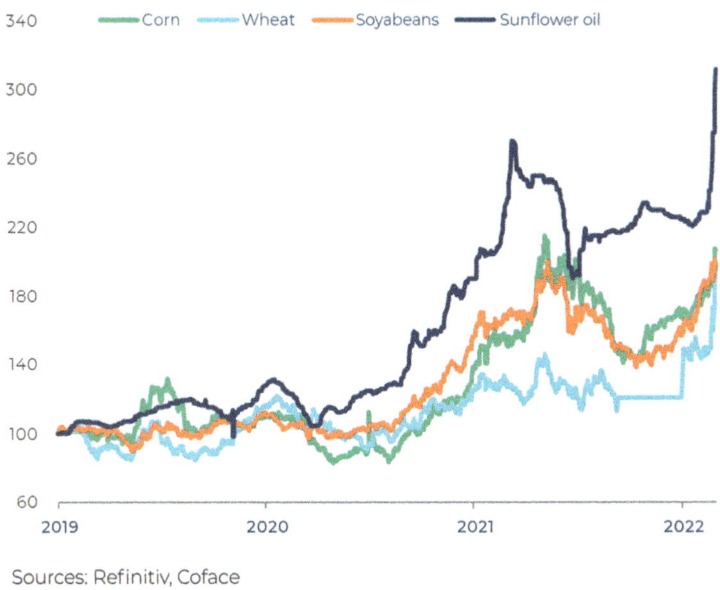

Fig. 1.3 Recent price changes of selected commodities

and waste management, both spatially and structurally. Home-based production and local production in close proximity to where people live have become excellent viable alternatives for dealing with the challenges ahead, including more secure sources of food, better nutrition, and income generation opportunities. But this structural transformation of production and consumption does not happen instantaneously; rather it takes time and major investments in infrastructure and productive facilities are required to sustain them.

Understanding and evaluating the impacts of the pandemic and other crises buffeting the world, is now at the center stage of coming to grips with their consequences, controlling the spread of the pandemic, and mitigating its negative impacts. In what follows we will highlight the salient characteristics of the pandemic, the wars in Ukraine and Gaza, and what to expect their specific impacts are going to be on the global economy and the specific impacts they will have on food systems in urban settings. The combined challenges of the pandemic and the war in Europe will be tackled at the same time.

What Shape Will the Recovery Take?

While it is true that the full impact of the virus is difficult to estimate at this time not knowing how long it would last and how many people it is likely to infect, there are enough economic and medical insights to inform our expectations and analysis of what is likely to happen, what are the likely costs of its manifestations, and what are the expected difficulties that we could face. Of course, the longer it takes to stop the escalation of the infection rates, the more people will get infected, the higher is the death rate, and the larger are the magnitudes of the resources needed to deal with it. Of course the faster policies and programs are implemented to mitigate its damages, the lower are the expected losses and their impacts. Dealing with the economic consequences of the pandemic and understanding them is also helpful in preparing the world to deal with other crises, specifically when seen against the onslaught of the conflicts in Ukraine, Gaza and of climate change.

What we know from simple economic analysis is that growth and contractions feed on themselves. Once economic contractions ensue, they become dynamically cumulative spreading out vertically (deeper) and horizontally (involving more sectors and regions). Historically, declines are steep and often get magnified as time passes and as more sectors and regions get affected.

So far a number of preliminary observations can be made about the likely manifestations, impacts, and outcomes of this stubborn pandemic. These observations are cumulatively developing so quickly that no one writer can collate them all or even capture a partial list.

First, the stock and money markets, as expected, reacted fast and steeply to the shocks and expectations of the decline driven by mass psychology of panic and customary overreactions. The US indices (Dow Jones, NASDAQ, and the S&P) suffered large losses, at one point, a decline of over 32% in less than a month. The Canadian indices (TSX, Venture and S&P) suffered equally large losses (over 29% loss in the TSX in the first 2 weeks of the initial shock), as did the London FTSE (with over 34% decline over the same short period) and the German GDAX with over 33% decline, whereas the Chinese HSI suffered losses of about 24%. Stock markets typically overshoot (exaggerate both the initial decline and growth) as stampedes develop and as investors hold unanimous expectations about the future.

Global stock markets erased gains garnered over 4 or 5 years in just a couple of weeks. It is difficult to believe that these losses, if they persist, would not have negative impacts on the real economy as they start to affect consumption and saving and investment decisions of consumers and businesses along the lines economists call Wealth Effects (or more technically Pigou Effects). Fortunately, most stock markets around the world recouped their initial losses under the weight of generous and large-scale quantitative easing and fiscal aid programs. If anything, stock markets exhibited bubbles behavior as corporations bought their own stocks and low interest rates, often negative real rates, supported stock values and drove the momentum for continued growth. Uncertainties about the war, inflation, and policy responses are again rattling asset markets.

Second, the economy is a set of interacting sectors and activities that typically propagate any contraction in any given sector to other sectors through indirect and induced effects. The simple fact is that the direct effects of contractions are poor estimates of the total and final effects. In summary, the direct consequences of the initial contractions breed indirect and induced contractions that sum to a large magnitude that could add to multiples of the direct effects.[9]

Third, the first round effects of the virus shock have been concentrated in the travel, accommodation, entertainment, sports, and restaurants sectors. But all other sectors of the economy were affected through the indirect and induced effects. However, not all of the sectors of the economy were affected in the same way, nor will all sectors contract. Surely the communications, software industry, internet use, online shopping, the home delivery industry, local food production, and pharmaceutical and cleaning materials have benefited from the spread of the virus and from people being locked at home and away from their work and family.

Fourth, one of the immediate consequences of the virus was the widespread questioning of the Global Economy's failure to deal with the Coronavirus, if not its role in propagating its negative effects. We are increasingly finding out that increased global travel has made the transfer of the virus that much easier, speedier, and more pervasive.[10] The globalized economy apparently has also created many empty economic boxes in many industrialized economies. Manufacturing in the advanced West was almost lost and outsourced. The USA did not have enough ventilators, masks, and gloves as these are no longer produced in America, but in low-cost labor regions and in countries with less stringent environmental and labor standards. Already there are serious calls for a thorough review of the increased risk exposure some countries have created for themselves due to this widespread redeployment of economic activity. Globalization, the way we know it, can no longer be taken for granted as people call for balancing its benefits against the inherent risks and challenges that global value and supply chains have faced in today's globalized economy. Surely, just-in-time medical systems, shortages of critical life-saving supplies, and food products are no longer safe bets and changes as to where they can be sourced are high on the policy agendas of many countries.

The need to reconsider and review globalization does not have to, however, come at the expense of multilateralism and international cooperation. There are inherent risks and costs to narrow nationalism and me-first policies.[11] This is a global crisis and calls for international cooperative strategies; we are all in it together. Individual, competitive procurements and solutions, and narrow nationalism would exacerbate the problem. Nonetheless, there are reasonable grounds to reassess the wisdom and costs of separating production and consumption activities especially of food and medicine. There is, however, a real cost to tampering and restricting globalization triggering higher inflation rates as countries opt to procure their needs from the high-cost domestic suppliers.

---

[9] The magnitude of the multiplier effects depends on leakages into savings, imports, and money.

[10] Statistics Canada, blames international travel for 27% of the infection cases in Canada. Statistics Canada, April 7, 2020.

[11] In March alone governments around the world introduced 70 new export curbs on medical supplies according to Simon Everett Professor of international trade at St. Gallen University in Switzerland (Jason Douglas, Wall Street Journal, April 4, 2010).

Fifth, a corollary of this globalized world is that any contraction in any part of the globe will travel to all other corners of the world. No country is now immune to contractions in China, Syria, or Italy. We are all in it together. More damaging perhaps is the fact that because we are all affected and possibly contracting together, the magnitude of losses will be magnified through the extensive trade channels that have emerged in the past four decades or so. Again not all countries would suffer equally from the impacts of the decline in world trade and incomes. The more trade exposed a country is, the larger are the impacts of the declines. The degree of exposure is measured by what economists call the Foreign Trade Percentage. This is the sum of imports and exports divided by GDP. This ratio in Canada is over 64.1% (2017), which is quite high. The trade to GDP ratio in the USA for the same year is much lower at 26.6%.[12] Trade increases our menu of choices and diversifies the products and services available to us but at the cost of increasing our vulnerability to factors outside our control. A balanced assessment of costs and benefits of trade, specialization and globalization is necessary to protect against rash policies that could cost us all the benefits of trade expansion.

Sixth, equally troublesome is that many poor countries in the world cannot fend for themselves. The Coronavirus has exposed the inequitable world we live in and the great divide between rich and poor countries. Poverty will travel, probably not as fast as this virus has traveled, but we will soon find out. Inequities in income and wealth distribution systems are not confined to countries but are manifested within countries. Rural economies have suffered more disproportionately than urban economies and labor bore more of the costs of the contractions than capital.

The ensuing economic demise will affect the poor and the vulnerable more drastically than the more fortunate. Poor people and the less fortunate among us live from paycheck to paycheck and have practically no savings to access in hard times. Actually, most poor households are already over indebted. Household debt in the USA and Canada are at an all-time high. A dominant characteristic of our capitalist system is the excessive polarization of the income and wealth distribution systems. The 1% versus the 99% is a glaring reminder of a major market failure we have allowed to rule our lives, where we find out that 300 billionaires have amassed more wealth than 50% of the people of the world.

Poverty is the best predictor of early illness, early hospitalizations, longer hospital stays, and earlier death.[13] It is also a reliable predictor of food insecurity. Poverty imposes a social cost on all of us as it compromises productivity gains by shaving 5.5–6.6% of productivity growth.[14] A more just distribution of especially wealth and income would mean fewer people would die from the virus and less will be in dire straits at the moment. This crisis has uncovered the need to reconsider the current system that produces and sustains these injustices and the feasibility and necessity

---

[12] Wikipedia, en.m.wikipedia.org.

[13] The Public Health Agency of Canada lists income and social status as the first socioeconomic determinants of health. (Second Report on the Health of Canadians. Government of Canada: Ottawa: Queen Printer, 1999).

[14] Hugh Segal. 2014.

of a Guaranteed Annual Income Scheme. A resilient economy based on a fairer distribution of income and wealth might not have exposed so many to the virus and might not have necessitated the draconian support programs that we now find so critically needed and costly. Paying the future costs of the present programs has resurrected the idea of a wealth tax or even an increase in the marginal tax rates on rich incomes. The top income bracket in the USA was in the low 90% range from 1944 to 1963, it is now 37%.[15]

Seventh, the current crisis has a sort of a silver lining as far as the environment is concerned. COVID-19 has put the brakes on a runaway economy. The pandemic has slammed on the brakes and jammed the GDP gearstick into reverse. In cities across the USA, Canada, China, and the world, traffic on roads and highways has fallen dramatically over the past few months as the Coronavirus outbreak has forced people to stay at home and everyday life has grounded to a halt. Pollution has dropped too and the earth has had a chance to have a respite. The pursuit of mindless and relentless growth is not sustainable; it comes now at the expense of survival of life on the planet. The COVID-19 crisis has given us a chance to reevaluate our damage to the environment and has given us a glimpse of what a clean environment would mean. It drove home the idea that mindless economic pursuits could threaten life and choices between economics and life may have to be made.

People in Venice have seen fish in the canals for the first time in decades and people in Shanghai could breathe a sigh of relief that have been denied to them as relentless economic growth has devoured the atmosphere. It makes sense to take a pause and reconsider the environment before we revert to our old ways of damaging the basis of sustainable life. On the negative side, the pandemic has spawned a major increase in medical waste and many countries are indiscriminately spraying harmful chemicals and disinfectants that will be harmful to the environment and would subtract from the positive environmental impacts the pandemic might have had.

Maybe it is time to pause and reconsider how we might preserve desirable economic objectives and redesign our economy in a sustainable way by choice and not by disaster (Peter Victor).[16] This is all the more important as Climate Change is likely to create crises similar to the one we are going through now. A greater emphasis on minimizing waste, reusing, recycling products, and emphasizing the circular economy in all aspects of consumption and production is emerging as necessary strategies for sustainability and survival on this planet.

Eighth, many programs are planned to make short-term benefits available to distressed workers and businesses. This is necessary and helpful. There is, however, an implicit danger that these demand-oriented policies may exacerbate the economic woes if they are not carefully designed and implemented. What we are contending with currently is a supply reduction compounded by a demand contraction. Typical macroeconomic stabilization policies are short-term demand-oriented policies; they

[15] https://taxfoundation.org/us-federal-individual-income-tax-rates-history-1913-2013-nominal-and-inflation-adjusted-brackets. And https://www.nerdwallet.com/blog/taxes/federal-income-tax-brackets/

[16] Peter A. Victor. Managing Without Growth: Slower by Design, not Disaster. New York, Elger. 2019.

are not well-suited to deal with supply shocks. The major economic problem we face today is the outcome of not only a demand shock and lost jobs and incomes, it is also a supply shock as factories, shops, and offices shut down, reducing supply. The war in Europe has exacerbated supply shocks. The last time we faced a supply shock problem was in 1974 in the aftermath of OPEC oil price shock. Costs of production increased and the aggregate supply of output decreased. That, by itself, would have reduced employment and raised prices. The reaction of governments to the problem was perhaps flawed as they tried to shore up demand to stave off increasing unemployment, which triggered a Stagflation problem of high unemployment and high inflation similar to what we suffered from in the mid-1970s and early 1980s.

We run the same risk now. Maintaining peoples' incomes during this crisis should be carefully administered to maintain the delicate balance between the demand for output and the supply of output. The longer this crisis lasts, the greater the dangers of upsetting this fragile balance. Already there are signs that inflationary pressures are building in many countries and these will spill over to many other countries across the globe. The US recent inflation data puts the rate at 8.6%, a rate that has not been seen for the past 40 years. The supply shortages of energy and food are already feeding dangerous inflationary expectations everywhere.

We need to make sure that the infection rate curve is flattened very fast for both epidemiological and economic considerations. In the meantime we may need nuanced supply-side policies using a few lessons from the industrial policies used by the Southeast Asian Tigers (South Korea, Singapore, Malaysia, and Hong Kong) and some European economies to shore up supply as demand policies are used to support aggregate demand. Supporting businesses to keep their workers, using spare hotel capacity to quarantine infected people or those suspected of being infected are examples of supply-boosting policies. I am sure entrepreneurs and policymakers can think of many more creative programs and measures.

Some who muse about the post-pandemic world (like Charles Goodhart and Manoj Pradhan, interviewed on March 23, 2021 as part of the 2021 Scholars' Webinar Series run by C.D. Howe Institute, focus on the long-run trends in aggregate supply and demand. They indicate that, in the decades immediately following WWII, there was rough balance between developments on the supply and demand sides. With urbanization, the shift from agriculture to manufacturing, and major changes in technology, our ability to produce increased rapidly. At the same time, the expansion of social programs and unions transferred spending power to those with higher spending propensities, and more young couples were setting up independent homes, so demand increased along with supply, and there was little upward or downward pressure on prices. In more recent decades, demographics have continued to be important, and globalization has become a central influence as well. The large baby-boom cohorts began saving for their retirement (reducing current demands for goods), and the emergence of China as a major source of produced goods increased supply. Further, the expansion in world trade complemented skill-biased technical change in lowering the incomes of less-skilled workers in the developed world, shifting spending power away from those who tend to spend more. But Charles and Manoj note that these deflationary developments were overridden by central bankers, who pursued expansionary policy (quantitative easing) to meet their (positive) inflation targets. With hindsight they may have over reacted and are now overreacting again trying to wrestle the inflation they, the pandemic, and the war have created.

There must remain similar uncertainty concerning the long-run structural influences on prices, since economists are not trained to predict politicians' decisions in this sphere either. We have some understanding about how long-run trends affect demand and supply (aging populations increase demand, less globalization decreases supply, addressing climate change raises investment spending in new technologies and so can eliminate secular stagnation). The policy choices are clear and what we do or don't do will affect the course of our economies in the post-COVID post-crisis world.

Ninth, what this crisis and the earlier one in 2008 have underscored is the pivotal role of governments in mitigating and moderating the economic impacts of these crises and challenges. What governments do and do not do will determine the depth of the crisis, its duration and its aftermaths. More importantly is the fact that the quality of government action and the governance system seem to matter a whole lot more than what people expected or could have expected in normal times in capitalist economies. The draconian measures, the unprecedented ones and those that would not have been expected, are proving to be crucial determinants of whether the crisis is controlled, subdued, moderated, or explodes into chaos. Government action in the face of this unprecedented, unanticipated, and profoundly damaging crisis has blurred the distinction between market-oriented governments and more interventionist governments. Pragmatic policies have trumped ideological differences. There remains, of course, the glaring difference between efficient, effective, and accountable governments and those governed by undemocratic government, populist leaders or demagogues.

Tenth, it is becoming clear that the longer the social distancing lasts there will be an erosion of social capital as people redefine their private spaces and social interactions. Actually, much trust in strangers and institutions may be lost already as a result of the crisis. This may, in turn, translate into a period of prolonged unwillingness of people to circulate in society (shop, dine out, socialize, see movies, etc.), contributing to extending the bottom of the U or L curve. This loss will be particularly acute if there is a second wave of infection following the initial period of distancing, something that may already be happening in many countries. People will feel especially vulnerable if, after an initial period of sincere effort to stop the virus' spread, it spreads again nonetheless.

Eleventh, the longer the lockdown lasts, the larger are the economic costs, the more pervasive are the disruptions, and the harder it will be to restart the economy.[17] Equally true is the fact that the longer the lockdowns last, the greater will be the conflict between epidemiology and economics. One of the most important questions right now is: How can we balance the risk of increased morbidity and death versus crippling the economy? Is there a way in which the economy can be revived with as little cost as possible to lives and health?

---

[17] Liam Casey, National Post, April 9, 2020. https://nationalpost.com/pmn/news-pmn/canada-news-pmn/dynamic-physical-distancing-could-help-balance-covid-19-fight-economy-study.

Already there are those who question whether the economic costs are now too high to sustain the social distancing and the economic closures. Increasingly economic considerations are raised to pressure political leaders to reconsider opening up in the economy using the argument that the economic costs are not only too high but also unwarranted given the "low death rates" and the Swedish example.

On the other hand, a few researchers are trying to see if the economy and epidemiology can be reconciled, where the infection rates and death rates can be minimized while allowing people to go to work. Or better yet that the situation can be controlled such that the economy can be restarted without affecting much the death rates that can be expected from closing the economy with social distancing. Two scientists from the University of Toronto and another from the University of Guelph used mathematical modeling to predict the course of the disease in Ontario under similar propositions.[18] Their base modeling found 56% of the province's population could become infected with the COVID-19 over the next 2 years. Their findings suggest that at the peak of the pandemic, 107,000 Ontarians would be hospitalized with 55,000 people in the ICU, should nothing be done. The province currently has about 2000 intensive care beds. They propose a novel dynamic physical procedure where off-and-on distancing could help keep the healthcare system from becoming overwhelmed while allowing periodic psychological and economic respite for populations. The key proposal is to peg physical distancing measures to the number of intensive care beds in use. These restrictions can be loosened when ICU-use numbers drop, then tightened up again, when the numbers near capacity.

The proposed dynamic modulation of off-and-on distancing raises a number of questions that remain; however, unanswered. Should the peg be the number of ICU beds or the number of deaths? Many of the patients that are transferred to ICU beds never come out alive, especially old people, minorities or people with medical issues, somebody's mother and father or grandparents. Second, how feasible is it to turn the economy on-and-off without disrupting its capacity for sustained performance? Third, what is the implicit trade-off here between dollars and life? What dollar value is put on a human life that is acceptable to politicians?

Of course increased testing, safety permits, and quarantining of the infected and their contacts may be necessary measures before this dynamic off-and-on system is implemented, but then if these complementary measures are instituted why do we need an on-and-off system, unless of course the medical system is overwhelmed.

How Has the Pandemic and the War in Europe Affected the World Food Systems?

The post-COVID and post-crisis world will be different than the pre-COVID pre-crisis one. The economy will recover, but it has been so far an uneven recovery, the

[18] https://www.medrxiv.org/content/10.1101/2020.03.24.20042705v1

inflationary pressures are gathering steam, the role of government in the economy is now pervasive and deep, and there are no signs that it will be reversed soon. Globalization is increasingly under scrutiny and there will be challenges to its pervasive role in the world economy. Just-in-time health systems will not be tolerated and empty national economic boxes, particularly in manufacturing, these will have to be filled. The efficiencies generated by increased specialization will be threatened as supply chains faltered and me-first policies prevail. The separation of food consumption from food production has provoked reevaluation and restructuring of food systems in the direction of increased reliance on local production and tying cities to their surroundings as sources of reliable food supplies. The world got a glimpse of what damage people have caused to the environment. Recovery will hopefully be increasingly anchored on green activities.

Nature-based Solutions (NbS) and strategies emerged in response to the need to prepare communities to deal with climate change challenges. They are very useful in redesigning food systems and agriculture to minimize their ecological footprint and to improve food security in the wake of the disruptions the COVID-19 pandemic and the war has wrought on communities and vulnerable groups. NbS are rooted in the understanding of nature and its natural processes, fostering the preservation of ecosystems' integrity, improvement of their sustainable management, and the creation of new ecosystems. At the heart of these initiatives is the preservation and enhancement of green infrastructure primarily in urban areas. This is not possible without the protection and preservation of green spaces whether through green roofs (Eco-Roof Incentive Program),[19] green streets, urban forestry, community gardens, green ravines, or stronger integration of the urban and peri-urban spaces in the urban food nexus. Canada alone has earmarked spending over $21 billion on green infrastructure in cities.[20] NbS investigate and demonstrate the role of urban green infrastructure as an often inexpensive way to simultaneously contribute to the mitigation and solving of heat island effects, flooding risks, and air quality problems while at the same time contributing to better-appreciated surroundings and food security, thus improving health and well-being. NbS are now complemented and harmonized with the circular economy; they tackle jointly community sustainable development and resilience to climate change.

A fundamental lesson-learned from the pandemic is that most of its problems are extensively connected to one vital sector of the economy: food. More than a third of the world's land is currently dedicated to food production. How we manage that land will impact levels of food security, greenhouse gases in the atmosphere, and whether plants, insects, and animals can thrive.[21] Although not mentioned directly, land without water is useless. When we mention land as a factor of production of food we typically mean the land–water nexus.

---

[19] https://www.toronto.ca/services-payments/water-environment/environmental-grants-incentives/green-your-roof/

[20] https://www.infrastructure.gc.ca/plan/gi-iv-eng.html

[21] FAO.2017. The Future of Food and Agriculture: Trends and Challenges. http://www.fao.org/3/i6583e/i6583e.pdf

The pandemic and the war have disrupted supply chains and communities found themselves competing for food that comes from thousands of miles away. Adding to climate change new concerns have been awakened by the COVID pandemic and now by the war and these have challenged countries and communities to reconsider their food systems within a risk management and environmental lens.

The existing food system has supported fast-growing populations and fueled economic development, but productivity gains have come at a cost. Today, 55% of the world population live in urban areas and by 2050, the urban population is expected to reach 68%.[22] Most of this increase in urbanization will be in low-income countries that explain over 90% of this increase.[23] This increase in the urban world population will occur primarily in Africa and Asia.

The growth in the urban population will generate an increasing demand for food, land, water, energy, and basic services. Today, cities already consume 70% of the global food supply and almost 80% percent of the total energy produced in the world, and they produce 70% of global waste.[24]

This growth did not come without problems. Many urban and peri-urban communities are exposed to food and nutrition insecurity and cities are overexploiting water resources. Many cities in the developing countries do not have wastewater treatment and appropriate sewage systems. Quite often this leads to pollution of the ground and surface water resources. Expanding cities are consuming land: large tracts of agriculture land are converted to urban development and industrial use. A major challenge today is dealing with the deforestation, linked to urban sprawl, which results in the loss of biodiversity and to watershed degradation, with a negative effect on water quality and in increase in landslides, floods, and water scarcity in the downstream areas.

Cities today are significant contributors to climate change, accounting for about 70% of global energy-related greenhouse gas emissions. Deforestation of agricultural land, as well as livestock and soil mismanagement, has saddled the food industry with the responsibility for almost 30% of greenhouse gas emissions globally, while mismanagement of fertilizers has caused eutrophication of waterways and chemical pesticide use has degraded the natural resources on which the food system depends.[25]

It is estimated that today for every dollar spent on food, two dollars' worth of health, environmental, and economic costs are created. Half of these costs—totaling USD 5.7 trillion each year globally—are the result of the way food is produced and how food waste and food by-products are handled.[26]

[22] FAO.2020. Green Cities Initiative: Green Cities Action Programme: Building Back Better.

[23] Ibid.

[24] Ibid.

[25] Tansy Robertson-Fall, https://medium.com/circulatenews/five-benefits-of-a-circular-economy-for-food-21c3654f4393

[26] https://foodprint.org/issues/the-problem-of-food-waste/

This need not be the case. The food system can be fundamentally redesigned and restructured to actively tackle climate change and biodiversity loss, promote human health, reduce food insecurity improve the resiliency of communities while reducing the overall costs of the system to society. This can be achieved by reconnecting communities with local food production, by changing the way we grow food, design food products, and handle by-products and waste. We can create a less wasteful system by promoting a circular economy for food that provides key benefits for society and the environment by improving nutritional health, reducing the ecological footprint, securing access, mitigating risks of supply defaults, and generating local incomes and jobs that support cities and local communities.

The Food Circular Economy and its Benefits

In a circular economy, food is produced in a regenerative mode. Regenerative food production involves the identification of a broad set of farming methods with two clear and complementary objectives: the production of high-quality food and the improvement of the surrounding natural ecosystem.

Employing farming techniques such as agroforestry and permaculture, the use of organic rather than synthetic fertilizers and employing activities such as rotational grazing, in which the management of animals boosts the productivity of the land, the targeting of crop rotation, where a range of crops are grown on the same piece of land across the growing seasons will pay handsome dividends in yield and sustainability. The emphasis of this system will be placed on diversity of crops that can provide health, environmental, and economic benefits.

Diversity of crops is crucial for health and resilience. Currently, more than 75% of our food comes from a mere 12 plant species and 5 animal species.[27] Dependence on only a small range of food types has led to concerns about diets being "energy-rich but nutrient-poor," with micronutrient deficiencies affecting at least 1.5 billion people. There are, however, over 200,000 edible plant species in the world, and growing a range of crops using regenerative methods is not only good for our health but also for nature.[28]

It is now documented that regenerative farming techniques build soil health enabling it to provide vital habitats for microorganisms at the start of the food chain—such as fungi and insects, and the animals that depend on them—and hold more water, improving fertility and productivity of the land. In Japan, farmers introduced ducks into rice paddies, as an alternative to chemical pesticides and fertilizers, where the ducks ate weeds and pests and deposited their waste on the fields. This practice improved soil fertility increasing rice yields by 20–50% compared to

[27] https://www.ellenmacarthurfoundation.org/explore/food-cities-the-circular-economy

[28] Tansy Robertson-Fall, https://medium.com/circulatenews/five-benefits-of-a-circular-economy-for-food-21c3654f4393

industrial rice plantations. In addition, the affected farms now also generate revenue from duck eggs and meat, as well as figs, vegetables, and wheat, highlighting that diversifying is also good for business.[29]

The Circular Economy in Food Production and Distribution Reduces Emissions

Planting cover crops that grow for long periods of time, employing agroforestry, and other regenerative methods that allow for photosynthesis to take place year-round, CO_2 is drawn from the atmosphere into the roots of plants and the soil microbes around them. At the same time, allowing roots and stubble from harvested plants to break down on the land returns carbon to the soil, and so does the use of organic fertilizers made from manure, composted plant by-products, and inedible food waste.

A circular economy for food that generalizes the use of these techniques and methods could reduce the sector's greenhouse gas emissions by 49%, or 5.6 billion tons of CO_2, by 2050.[30] Organic waste flows from towns and cities can be valorized. Each year cities generate more than 2.8 billion tons of organic waste, with only 2% of this currently looped back into the system for use. Recognizing the potential to use this organic matter to store more carbon in farmland, businesses, governments, and other organizations are turning to anaerobic digesters to turn organic waste into compost. From small projects, such as that in Malton, UK, to larger projects like Plant Chicago, USA, potential gains are beginning to be realized.[31]

Elsewhere, businesses are aiming to reduce greenhouse gas emissions by designing out edible food waste. Reducing edible food waste by only 50% by 2030 has the potential to reduce annual emissions by 1.4 billion tons of CO_2 and a number of businesses are already focused on this goal. Turning what would typically be food by-products, including corn cobs, edamame pods, and beetroot skins, into food products in Japan. In the Netherlands, Surplus Food Factory uses the tops and bottoms of tomatoes not wanted by hamburger producers to make soup, along with other food items that are considered unsuitable for commercial use elsewhere. Chefs, food product developers, and others responsible for designing the products and food that we eat can all contribute to reducing emissions by making more of food. In Canada and the USA food waste often exceeds 50% of food purchased. The potential of savings focused on food waste reductions are substantial.[32]

[29] Tansy Robertson-Fall, https://medium.com/circulatenews/five-benefits-of-a-circular-economy-for-food-21c3654f4393

[30] Ibid.,

[31] Ibid.,

[32] https://foodprint.org/issues/the-problem-of-food-waste/

Reconnecting Urban Centers with Their Peri-urban Surroundings Is the Right Strategy

By reconnecting cities with surrounding peri-urban areas, a circular economy for food can build resilience in the food system, improve food security, save money, reduce emissions, and create jobs. More than 40% of the world's irrigated cropland is located in peri-urban areas, yet the food produced on this land is often flown to consumers on the other side of the world, while similar products are imported into neighboring cities.[33] When global supply chains are disrupted, by events like the COVID-19 pandemic, access to food is jeopardized. But by reconnecting towns and cities with local food production and balancing this with global supply, resilience can be built into the food supply chain so that no matter where or what disruption is faced, food can still be accessed by those who need it. Redistributing surplus edible food could feed one billion people by 2050. Paris city authorities have mandated a policy that demands all of the arrondissement must reconfigure their supply chains in such a way as to ensure that most of the provisions residents purchase should be available within at most a 15 minutes journey.[34]

In Medellin, Colombia, the Huertas para el Abastecimiento (Gardens for Canteens) program allowed for 20 tons of food in local areas to be sourced during the first 2 weeks of the City's COVID19 pandemic lockdown.[35] Local producers were supported by the Council to distribute their produce through private companies, facilitating supply to popular canteens and ultimately feeding the most vulnerable populations in the city. Similar decentralization of supply was successful in supporting communities in India, Kenya, and Vietnam.[36]

In China, the UK, Canada, and the USA, community-supported agriculture (CSA) sales and subscriptions have seen significant growth during the pandemic. CSA connects food producers to each other and to customers, establishing local food networks. In doing this, they provide food security to local communities while also providing smallholder farmers with guaranteed income.

Building Resilience in Local Communities

Around 70% of all the food we eat comes from smallholder farms, of which there are approximately half a billion globally.[37] As countries become more affluent, more often than not agriculture is industrialized, which threatens the livelihoods of

[33] https://www.ellenmacarthurfoundation.org/explore/food-cities-the-circular-economy

[34] https://www.parisdiscoveryguide.com/paris-arrondissements.html

[35] https://www.acimedellin.org/antioquia-le-propone-a-colombia-el-modelo-de-huertas-urbanas-como-solucion-a-la-malnutricion-en-las-ciudades/

[36] https://www.ellenmacarthurfoundation.org/explore/food-cities-the-circular-economy

[37] http://www.fao.org/3/i6583e/i6583e.pdf

smallholder farmers with possible knock-on effects including depopulation of rural areas. By connecting people with local food production, however, smallholder farms can be preserved, benefitting both their communities and the environment.

Despite providing such a large amount of our food, these farms only use 30% of resources and 12% of the world's agricultural land. Acknowledging the benefits for both the environment and people, the city of São Paulo has put in place mechanisms that enable the local community of farmers to thrive while providing those in need with locally grown food. To accomplish this, the city has devoted 30% of its territory to smallholder agriculture and City Hall sources food for school meals and food banks from this area to ensure that those in need have access to fresh fruits and vegetables. The initiative also secures livelihoods for farmers and their families, with the added benefit that the green belt around the city contributes to cleaner air. Toronto in Canada has embarked on expanding and allocating green spaces for community farming and this program is becoming a standard program in many Canadian cities.[38]

Urban–rural relationships like this are increasingly being seen as a priority to build resilience into communities of all sizes. The UN Food and Agriculture Organization (FAO) recently launched the Green Cities Initiative, which aims "to increase people's well-being," with beneficiaries including both city dwellers and "stakeholders related to green spaces."[39] Outlined in the Initiative's action program, activities will include connecting producers with local markets, as well as creating innovative agro-processing food hubs and "green" jobs, and advocating for use of local products. Many of the initial target cities are situated in developing regions where developing a circular economy for food could enable smallholder farmers to leapfrog industrial food production.[40]

The Circular Economy Creates Value, Incomes, and Jobs as it Saves Money

By sourcing food grown using regenerative and locally where appropriate, and making the most of food and food waste, cities could generate benefits worth USD 2.7 trillion a year by 2050.[41] The city of Brussels, for example, stands to gain over USD 130 million per year by producing 30% of its food locally using regenerative

[38] https://www.toronto.ca/services-payments/water-environment/environmental-grants-incentives/green-your-roof/

[39] FAO (2020).

[40] https://www.sitra.fi/en/publications/cities-circular-economy-food/#source-food-grown-regeneratively-and-locally-where-appropriate

[41] https://youmatter.world/en/food-system-broken-fix-it-27849/

practices, halving food waste, and converting half of all remaining organic waste into high quality compost.

Harnessing organic waste streams, which contain high levels of nitrogen and phosphorus, alone could unlock upward of USD 700 billion for cities globally, and valorizing these streams is already creating new business opportunities. Since 2008, UK-based Agriprotein has been turning food waste into products, including oil, organic fertilizer, and animal feed. Each of its facilities is able to process at least 350 tons of food waste per day, which is then fed to fly larvae. The larvae are then harvested to create Agriprotein products. In Kenya, Sanergy similarly creates insect-based animal feed and organic fertilizer, this time through collection of human waste. Sanergy provides residents of urban slums with no-sewers–based sanitation solutions and franchises them to local community members, collecting food waste each day to be turned into useful products.

Increasing use of organic fertilizers rather than chemical alternatives also yields health benefits, with the potential to reduce health costs by USD 550 billion globally.[42] Farm workers' exposure to pesticides currently costs almost a trillion US dollars and long-term exposure to low levels of pesticides has been linked to cancer, asthma, depression, reduced IQ, and higher rates of attention deficit hyperactivity disorder (ADHD)—the last two alone are costing the EU an estimated USD 150 billion each year. As the economic benefits of a circular economy for food become clearer and more businesses and policymakers will invest in developing the relationships that allow for nutrients to continuously be cycled, the system will begin to work better for all—working with nature to safeguard biodiversity, promote human health, and stabilize our climate.

Conclusion

In many respects the pandemic, the war in Europe and the new ones that may arise as climate change impacts intensify are unique being abrupt, global, of uncertain magnitude and duration, and necessitating draconian measures that exact a heavy toll on the economy and society. More important perhaps are the gaps that the many crises have uncovered in the economic, health, global supply chains, food systems, and government systems. Globalization has brought the world tightly together and computer and information technologies have compressed time and space. However, it is difficult to sing the praise of a system that left the most powerful and the rich economies unable to produce sufficient masks and ventilators to treat their sick. But this is not an acceptable argument to shun multilateralism and international cooperation. Now more than ever collective and cooperative action is needed to defeat the global crises.

[42] https://www.linkedin.com/pulse/meet-future-your-plate-circular-agriculture-andrei-geica/

The huge divide between the rich and the poor chipped away at the resilience of the economy and its capacity to sustain people. Large segments of the population are in dire need for support, particularly those with low incomes who are living from paycheck to paycheck. The crises have also given us a glimpse of what a clean environment would look like and how the relentless pursuit of growth has congested cities, choked the lungs of people, and threatened existence itself. They also engendered local food production and tying local food consumption to local food production in the surrounding areas and the need to address inequality of opportunities and outcomes at the local and the global levels. In more than one way these crises are underscoring the need for policies and strategies that can bring about better environment, better production, better nutrition, and a better life for everyone.[43] The aim is to ensure sustainable consumption and production patterns, through efficient and inclusive food and agriculture supply chains at local, regional, and global level, thus ensuring resilient and sustainable agri-food systems in a world prone to existential crises. The issue of emphasizing better is an issue of survival and is grounded in necessity and the pursuit of the best.

There are many among us who are hoping to see this crisis blown away and to return to the same old ways. That would be unfortunate as we would lose a golden opportunity to rethink and reconsider our economies, social systems, health systems, global systems, food systems, good governance, and political leadership. There are many needed changes to consider and ponder about how to increase our resilience in facing future challenges and to reconstruct a more equitable and more sustainable world. If there is a lesson to learn from the pandemic crisis and the conflicts in the Ukraine and Gaza, it is that their negative consequences are likely to be repeated with the onslaught of climate change. Now it is high time and an opportune moment to seek better environments, better production, better nutrition, and a better life. UNFAO has made the call for not leaving anyone behind, our survival is now at stake.

References

FAO (2020). Green Cities Initiative: Green Cities Action Programme: building back better.

Global Hunger Index (GHI). (2020). One decade to zero hunger, linking health and sustainable food systems, https://www.globalhungerindex.org/.

Ma, L., Nyuk, W. P., Soon, C. F., Hassim, M. F. N., Misbah, S., Rahmat, Z., Yong, W. T. L., & Sonne, C. (2021). Covid-19 pandemic in the lens of food safety and security. *Environmental Research, 193*, 110405. https://www.sciencedirect.com/science/article/pii/S0013935120313025

[43] UNFAO Strategic Framework 20,022–2031. https://www.fao.org/strategic-framework/en

Chapter 2
Redesigning and Transforming Food Systems to Be More Resilient to Disruption and Disaster: Lessons Learned from COVID

Nathaniel K. Newlands and Amani Alfarra

Abstract The COVID global emergency provides an opportunity for redesign and transformation in the agricultural sector—whether in a rural, peri-urban, or urban environment—for building resilience and sustainability in the face of many systemic challenges. These include climate change and extreme weather events, biodiversity, pest and disease, water availability and quality, livelihoods, population, health, and urbanization. Ecosystem-based solutions (EbS) offer an integrated or nexus approach to enhancing the resilience and sustainability of agroecosystems, but require strong engagement and coordination of stakeholders to codesign, pilot, and evaluate strategies and actionable solutions to be effective. Digital agriculture can help with the design, evaluation, assessment, and implementation of EbS using big data, predictive analytics, and Internet-of-Things, to help manage agroecosystem resources more efficiently and effectively.

Keywords Agroecosystem · COVID · Digital agriculture · Ecosystem-based solutions · Food security · Internet-of-Things · Nexus · Resilience · Sustainable Development Goals

---

N. K. Newlands (✉)
Agriculture and Agri-Food Canada, Summerland Research and Development Centre, Summerland, BC, Canada

Department of Geography, University of Victoria, Greater Victoria, BC, Canada
e-mail: nathaniel.newlands@agr.gc.ca

A. Alfarra
Land and Water (NSL) Division, Food and Agriculture Organization of the United Nations, Viale delle Terme di Caracalla, Rome, Italy

Walter G. Booth School of Engineering Practice and Technology, Hamilton, ON, Canada
e-mail: Amani.Alfarra@fao.org

A. A. Kubursi et al. (eds.), *Urban Food Security in a Crisis Prone World*,
https://doi.org/10.1007/978-3-031-89440-4_2

Introduction

The COVID pandemic is a timely reminder of how foreseen and unforeseen hazards within the complex and changing global risk landscape can affect lives, livelihoods, and health.[1] In 2020, nearly 2.37 billion people did not have adequate food, increasing by 320 million people since 2019 over the course of the COVID pandemic[2] (Boero et al., 2020). Food insecurity is also prevalent across Canada; before the pandemic in 2017/18, 12.7% of surveyed households (103,500) from the Statistics Canada's Canadian Community Health Survey (CCHS) experienced food insecurity, with the lowest prevalence in Quebec and highest in Canada's northern communities (e.g., Nunavut, Northwest Territories). During the pandemic, food insecurity in Canada[3] had increased to 14.6% or 1 in 7 households in May 2020 based on a survey of negative food experiences[4] (Deaton & Deaton, 2020). The First Nations Regional Health Survey (RHS) has further determined that 50.8% of households living in First Nations communities are food insecure (First Nations Information Governance Centre, 2018). Without government support measures such as the Canadian Emergency Response Benefit (CERB) and other recovery benefits (launched in September 2020 and ended in October 23, 2021) to counteract the pandemic's impacts on household disposable income, food insecurity would likely have increased far worse (Deaton & Deaton, 2021). COVID has shown us just how fragile and interdependent our food systems are, with many unexamined and vulnerable to collapse. From partial to full lockdown measures and other mobility and access restrictions, COVID measures have led to profound changes in consumers buying behavior and food consumption patterns, disturbances in transportation networks, and the closure of some food suppliers (Lioutas & Charatsari, 2021).

Food insecurity is complex, with countries interdependent on food imports and exports to varying degrees, with their situations changing rapidly during times of emergencies and disasters. In the case of Canada, agriculture and agri-food production relies on being able to maintain and expand its exports in international markets. In 2020, Canada exported nearly $74 billion in agriculture and food products (includes raw agricultural materials, fish and seafood, processed foods), increasing from $67 billion in 2019.[5] In 2020, exports supported a total production value of

---

[1] United Nations (UN). 2020. Hazard definition and classification review—technical report. United Nations Office of Disaster Risk Reduction (UNDRR), Geneva. 88 pp. https://council.science/wp-content/uploads/2020/06/UNDRR_Hazard-Report_DIGITAL.pdf

[2] FAO. 2021a. The State of Food and Agriculture 2021. Making agri-food systems more resilient to shocks and stresses. Rome, FAO. https://doi.org/10.4060/cb4476en

[3] Statistics Canada. 2020. Food insecurity during the COVID-19 pandemic, May 2020. https://www150.statcan.gc.ca/n1/pub/45-28-0001/2020001/article/00039-eng.htm

[4] Tarasuk V. and Mitchell A. 2020. Household food insecurity in Canada, 2017–18. Toronto: Research to identify policy options to reduce food insecurity (PROOF). https://proof.utoronto.ca/resources/proof-annual-reports/household-food-insecurity-in-canada-2017-2018/

[5] AAFC 2020a. Overview of Canada's Agriculture and Agri-Food Sector, Available online (Accessed3/14/2022):https://agriculture.canada.ca/en/canadas-agriculture-sectors/overview-canadas-agriculture-and-agri-food-sector

$139 billion or 7.4% of Canada's gross domestic product (GDP).[6] Despite rising food security, the value of its exports increased by 10% (2020 versus 2019), despite many challenges related to COVID and extreme weather events.[7] Food systems (FS) are highly integrated between Canada and the USA relying on vertically integrated food supply chains (FSCs), which leverage large-scale production, streamlined operations, and centralized planning and control to provide consumers with a consistent, timely supply of food (Marusak et al., 2021). Post-COVID, FS may undergo a broader transformation in global supply-chain specialization than they have previously for "just-in-time" delivery, involving structural changes in actions and firm consolidations, product diversification, extended food reserve capacity, and greater automation and use of digital agricultural technologies such as artificial intelligence (AI) and block-chain (Weersink et al., 2021). This could lead to more growth and reliance on regionalized food supply chains (RFSCs) that have, so far during the COVID pandemic, been sufficiently adaptive and responsive to changes in demand and delivery requirements, quickly pivoting to distribute products directly to consumers safely.

Water is a critical input for *agricultural* production and plays an important role in food security, so managing water sustainably is a key to the future of agriculture and achieving climate change and global Sustainable Development Goals (SDGs). Agriculture is not only a major contributor of greenhouse gases (GHGs), but also of water use, water quality degradation, and habitat loss globally (Willett et al., 2019). Agriculture, directly and indirectly, also contributes to water pollution due to fertilizer run-off, livestock effluents, pesticide, and other harmful contaminants. FS are highly vulnerable to climate change impacts —namely, hydrological-related risks such as floods and droughts (Holleman et al., 2020). Climate models show that Canada's agricultural regions will likely see drier summers from coast to coast, but increased winter and spring precipitation.[8] Irrigated agriculture represents 20% of the total cultivated land, and requires 70% of water use worldwide (over 40% in many OECD countries), yet contributes only 40% of the total food produced worldwide.[9] Agriculture is expected to face increasing competition from nonagricultural water demands due to rising urban population density and the energy and industry sectors, with an estimated 50% increase in urban water demands over the next 30 years.

The COVID pandemic and climate change are both global, complex crises—events or changes causing disruptive, damaging, and detrimental effects on social

[6] AAFC 2020b. Agriculture and Agri-Food Canada Departmental Results Report 2019–2020. Report No. A1-30E-PDF. Government of Canada. Available online (Accessed 3/14/2022): https://agriculture.canada.ca/sites/default/files/legacy/pack/doc/drr-rrm_2019-20-eng.pdf

[7] AAFC. 2021a. Departmental Plan 2021–22. https://agriculture.canada.ca/sites/default/files/legacy/pack/pdf/dp-pm_2021-22-eng.pdf

[8] Climate Atlas of Canada (Agriculture and Climate Change): https://climateatlas.ca/agriculture-and-climate-change

[9] The World Bank. Water in Agriculture, Online: https://www.worldbank.org/en/topic/water-in-agriculture#1

economic environmental systems with highly risky consequences and potential impacts that are difficult to anticipate or forecast (Lioutas & Charatsari, 2021 and references therein). As the COVID crisis is reshaping our world, so is the climate change crisis. Under the 2015 Paris Agreement on Climate Change, 196 countries committed to limiting global warming to well below 2 °C (i.e., below 1.5 °C) compared to preindustrial times. To reach 1.5 °C, the world must cut emission to 45% below 2010 levels by 2030, and reach net-zero by 2050.[10] Net-zero refers to the balance between the amount of GHG emissions produced and the amount removed from the atmosphere and is achieved when the amount added is no more than the amount able to be removed. The COVID crisis we are facing today has much to teach us about climate change, as arguably, a greater existential threat. Key lessons learned from COVID that apply to grappling with climate change are: (1) there is no time to wait (at the current rate of global greenhouse gas emissions, the 1.5 °C target will be missed leading toward a 3-4 °C rise in temperature); (2) Cooperation on intervention and adaptation strengthens resiliency decision-making; (3) Adversity prompts innovation for slowing down climate change; (4) digital data, knowledge, and technology are critical for ensuring society is connected, operational,and resilient during crises and disasters; and last, but not least: (5) the importance of building, strengthening, and maintaining transparency and trust across all levels of the food system, between national and local governments, companies and citizens.[11]

Complex Food Systems

A food system (FS) encompasses all interlinked processes, people, and infrastructure, and value-adding activities involved in feeding a population from growing, harvesting, processing, packaging, transporting, marketing, consumption, to the distribution and disposal of food and associated materials. A *sustainable* FS provides food security and nutrition for all in such a way that the economic, social, and environmental bases to generate food security and nutrition for future generations are not compromised. Global food demand is expected to increase between 59% and 98% by 2050, as the human population reaches an estimated 9.7 billion people, growing 83 million annually or 1.1% per year to 10.9 billion by 2100, before

---

[10] IPCC, 2021. Masson-Delmotte, V., P. Zhai, A. Pirani, S. L. Connors, C. Péan, S. Berger, N. Caud, Y. Chen, L. Goldfarb, M. I. Gomis, M. Huang, K. Leitzell, E. Lonnoy, J.B.R. Matthews, T. K. Maycock, T. Waterfield, O. Yelekçi, R. Yu and B. Zhou (eds.) Summary for Policymakers. In: Climate Change 2021: The Physical Science Basis. Contribution of Working Group I to the Sixth Assessment Report of the Intergovernmental Panel on Climate Change (IPCC). Cambridge University Press,https://www.ipcc.ch/report/ar6/wg1/downloads/report/IPCC_AR6_WGI_Full_Report.pdf

[11] Adapted from: DiBianca, S. 2020. Climate change and pandemic: 5 things COVID-19 has taught us about climate crisis. Salesforce: News & Insights. Published online. https://www.salesforce.com/news/stories/covid-19-and-the-climate-crisis-why-its-more-important-than-ever-to-prioritize-the-planet/

plateauing due to declining global fertility rate.[12] Population growth is expected to occur mostly in Africa, and abate in the Americas, Europe, and parts of Asia, especially as families in more developed nations have fewer children.[13] Socioeconomic migration in response to climate impacts may range from mobility as a proactive adaptation strategy to forced displacement in the face of life-threatening risks. It could lead to nearly 3% of the population (totaling more than 143 million people) in three regions—Sub-Saharan Africa, South Asia, and Latin America—to move within their country of origin by 2050.[14] FS shocks and disruptions circulate around the globe direct supplier shortages of specific food types and indirectly through international market trade. Shocks are also amplified by societal responses such as panic buying and food export bans that drive food price increases on international markets (Challinor & Benton, 2021). Hunger, malnutrition, and the unaffordability of a healthy diet is driven, however, not only by food unavailability, but by persistent poverty and income inequality. When FS fail, the resulting disorder threatens our education, health and economy, human rights, peace, and security.

FS that dominates today, termed dominant FS, were designed for the mass production of inexpensive, standardized foods. Their design focused primarily on intensification (i.e., productivity enhancement) without integrating social-economic and environmental sustainability. There is significant, broad evidence that supports the need for a major transition from dominant FS to sustainable FS that are more complex, and integrate broader sustainability development principles toward global SDG goals (Moberg et al., 2021; Pretty, 2020; Pretty et al., 2018). Sustainable FS can exist in many forms depending on their degree of efficiency, substitution, and redesign. Efficiency refers to making better use of on-farm and imported resources. Substitution is the replacement of technologies and practices into sustainable ones. Redesign refers to the scaling up of processes, technology, and practices to geographic scales beyond a local landscape scale (Vanbergen et al., 2020). Intensification of production in sustainable FS, unlike current dominant FS, requires increasing agricultural production without adverse environmental impact (i.e., sustainable intensification, SI) and without the conversion of additional nonagricultural land. At the farm-scale, SI may integrate conservation agriculture involving zero-tillage and use of cover crops to avoid fallow land. It may include shelterbelts/trees, pasture, and forage, intensive small and patch systems, integrated crop, pest, and biodiversity management, and irrigation water management. One-third of all farms

[12] UN, 2019. Department of Economic and Social Affairs, Population Division (2019). World Population Prospects 2019: Highlights (ST/ESA/SER.A/423). https://population.un.org/wpp/Publications/Files/WPP2019_Highlights.pdf

[13] Cilluffo, A. and Ruiz, N.G. 2019. World's population is projected to nearly stop growing by the end of the century. Pew Research Centre. https://www.pewresearch.org/fact-tank/2019/06/17/worlds-population-is-projected-to-nearly-stop-growing-by-the-end-of-the-century/

[14] The White House. 2021. Report on the Impact of Climate Change on Migration. Washington, D.C., USA. Available online (Accessed 3/14/2022): https://www.whitehouse.gov/wp-content/uploads/2021/10/Report-on-the-Impact-of-Climate-Change-on-Migration.pdf

worldwide have now crossed a redesign threshold, practicing forms of SI, and approaching a tipping point (Pretty et al., 2018).

Food System Transformation and Redesign

The COVID global emergency provides an opportunity for redesign and transformation in the agricultural sector—whether in a rural, peri-urban, or urban environment—for building resilience and sustainability in the face of many systemic challenges. These include climate change and extreme weather events, biodiversity, livestock/crop pest and disease, water availability and quality, livelihoods, population, health, and urbanization. Klerkx and Begemann (2020) argue that to understand how FS transformation is enabled in terms of forces, catalysts, reduced barriers, a mission-oriented agricultural innovation systems (MAIS) approach is needed to link innovation systems related to transformative concepts (e.g., agroecology, digital agriculture, Agriculture 4.0, etc.) across geographical scale. With growing consensus and embracing of a multiscale, action-oriented approach by policymakers, stakeholders, and scientists, different country-level typologies that have been developed for tracking progress toward sustainable FS. Nonetheless, it is realized that FS are inherently dynamics with many different types existing synchronously and transforming over time. One example is the FAO's Progress toward Sustainable Agriculture initiative (PROSA) framework, which comprises 16 indicators developed based on FAOSTAT data (1961–2018). Tubiello et al. (2021) summarize recent progress toward sustainable agriculture across four food systems production typologies that integrate food supply chains and food environment considerations. Four main topologies are defined, namely: Traditional, Land-intensive and Capital-intensive Mixed Systems, and Modern. These topologies are defined in Section 1.4 and synthesized in Table 3 (page 37) of the High-Level Panel on Food Security and Nutrition (HLPE)'s conceptual framework of the United Nations Committee on World Food Security[15] (HLPE, 2017). Traditional FS are smallholder farms that produce food locally and seasonally. Mixed FS involve smallholder and larger, distant farms providing greater availability to foods outside of typical production seasons. Modern FS involve smaller to industrial-sized farms producing a wide array of foods globally, making food available anywhere, anytime, and rely on demand from urban consumers living far from where food is produced. The Food Systems Dashboard Framework, launched in 2020, has further adapted and extended the HLPE's conceptual framework[16] to include 166 indicators from 20 data sources, many open-source, identifying 42 actions for orienting FS to be more sustainable

[15] High-Level Panel of Experts (HLPE). 2017. Nutrition and food systems. A report by the High Level Panel of Experts on Food Security and Nutrition of the Committee on World Food Security, Rome.

[16] Food Systems Dashboard Framework, https://foodsystemsdashboard.org/data-sources-and-methodology

(Fanzo et al., 2020). It is evergreen, being updated continually with new data, new indicators, and subnational data when available. It serves as both a policy decision-maker, and regional-to-local community resource enabling a more complete, shared understanding of FS and the challenges faced, actions that need to be prioritized and decided to improve diets and nutrition, and ultimately both human and environmental health, in tandem. Accelerating the transition of more FS around the globe requires the codesign of scientific and farmer input into digital technologies and sustainable management practices.

Redesign efforts need to move beyond the most common single activity of planting trees and paired activity of crop and biodiversity redesign (i.e., integrated pest management, conservation agriculture and soil health, agro-forestry, and irrigation management) in FS. Wider adoption and broader integration of watershed management, nutrient cycling, integrated pest management, and conservation agriculture for soil health is needed. Unintentional consequences and changing risks of climate change, new crops, SI practices on plant health, ecological services, and biodiversity need to be assessed proactively, not reactively. This will provide greater foresight for identifying and mitigating potential threats, before they impact FS. Policy-based incentives can be designed to better enable synergistic benefits across whole landscapes and are not just targeting environmental outcomes, but also payments for SI-based production. Overall, sustainability transition will require public policy leadership and private-sector technological innovation alongside consumers who culturally value and can afford healthy, sustainable diets (Moberg et al. 2021) and increasing awareness and support from people across rural, peri-urban, and urban foodscapes. With the aim of enhancing the capacity to prevent, anticipate, absorb, adapt, and transform in the face of risks due to COVID, climate change, and other anticipated/unanticipated shocks and stresses, a resiliency framework specifically for FS has been recently proposed.[17] This resiliency framework could be integrated by countries into their national disaster risk reduction (DDR) strategies that also comprise a systematic, whole-of-society approach for identifying, assessing, and analyzing the causal effects of disasters and reducing the risks and impacts of disaster based on risk assessments. Integrating lessons-learned from COVID, extreme events, and disasters to inform and guide agricultural FS resiliency planning would help to spur and accelerate actions to adopt measures based on an evidence-based approach for strengthening risk governance and management, resilience investment, and enhanced disaster risk preparedness for effective response and recovery at all levels (Bojić, et al. 2019; Savary et al., 2020). This is required to meet the global goals of the Sendai Framework for Disaster Risk Reduction (2015–2030).[18]

---

[17] FAO, IFAD, UNICEF, WFP and WHO. 2021a. The State of Food Security and Nutrition in the World 2021. Transforming food systems for food security, improved nutrition and affordable healthy diets for all. Rome, FAO. https://doi.org/10.4060/cb4474en

[18] Sendai Framework for Disaster Risk Reduction https://www.preventionweb.net/files/44983_sendaiframeworkchart.pdf

Urbanization is increasing, with 55% of the world's population (projected to reach 8.5 billion by 2030) residing in urban areas that consume 70% of global food supply. Urban agriculture (UA) is practiced by an estimated 800 million people worldwide (Pretty et al., 2018). UA refers to cultivating, processing, and distributing food such as animal husbandry, aquaculture, urban beekeeping, and horticulture. UA can also address rising water demands and alleviate food shortages and enhance self-sufficiency in urban areas involving non-potable water reuse for crop irrigation, groundwater replenishment, and industrial processes. An urban area is defined as a human settlement with high density and built infrastructure. While not yet amenable to a standard definition,[19] in Canada, it is defined[20] as 1000 persons with density greater than 400 persons/km^2. UA is interlinked with Green Infrastructure (GI), defined by the US EPA (2012) as, "systems that use vegetation, soils, and natural processes to manage storm water and generate healthier urban environments," and US Water Infrastructure Improvement Act (2019) as, "the range of measures that use plant or soil systems, permeable pavement or other permeable surfaces or substrates, storm water harvest and reuse, or landscaping to store, infiltrate, or storm water, reducing flows to sewer systems or to surface waters." GI disconnects impervious surfaces from gray infrastructure such as dams, seawalls, roads, pipes, and water treatment plants, to provide flood protection, cleaner water, and cleaner air. GI can be applied at both the local/building and regional/landscape scales and installed indoors and outdoors. Locally, GI includes rain and community, rain, rooftop gardens or "green roofs," indoor vertical farms, hydroponics, aquaponics, infiltration planters, tree boxes permeable pavements, infiltration planters, trees, and rainwater harvesting systems. At the landscape scale, it includes the preservation and restoration of wetlands, floodplains, forests. Ecosystem- or Nature-based solutions (EbS/NbS) (Sonneveld et al., 2018) and proactive and precautionary, biovigilance strategies are aimed at enhancing resilience and sustainability of FS, while at the same time anticipating and reducing ecosystem risks[21] (Carisse et al., 2017). Such integrated approaches offer novel, integrated approaches for monitoring, assessing, and co-learning from across multiple actions and interventions.

---

[19] UN Statistics Division, https://unstats.un.org/unsd/demographic/sconcerns/densurb/densurbmethods.htm

[20] Statistics Canada, https://www.statcan.gc.ca/en/subjects/standard/pcrac/2016/introduction

[21] AAFC 2021b. Biovigilance: A modern approach to plant pest management. Vision Statement. National Biovigilance Working Group. Unpublished. 32 pp. (See: Establishing a Plant Health Science Vision for Canada: https://inspection.canada.ca/about-cfia/transparency/consultations-and-engagement/partnerships-pahs/pahs/eng/1561741379774/1561741465925

Digital Agriculture Solutions

New evidence-based performance metrics, data-driven scientific methods, and decision-making tools are, however, needed to provide guidance, encourage adoption, and mitigate risks to put such integrated approaches into use and action (Newlands, 2016). Increasingly, science is becoming multidisciplinary with traditional disciplines breaking down or "de-siloing," alongside policy and implementation decision-making moving away from a sector-based approach to a more holistic, ecosystem-based approach. This increases the need for examining trade-offs and sustainability performance at different scales and levels of complexity in ecosystems, requiring digital agriculture—the integration of big data (BD), and artificial or augmented intelligence (AI), and general equilibrium (GA) or "whole sector or economy" optimization to unravel such complexity in data, models, and decision-making (Ayed & Hanana, 2021; Newlands, 2016). Large, complex datasets that require a high degree of automation and computers for processing and analysis (i.e., machine-readable data) is termed "Big data." This includes large, raw, or unstructured data being collected at rapid rates challenging the use of conventional relational database and techniques (Newlands, 2021). Decision-making tools and other applications that harness big data can increase farmers' decision-making capacity and is a valuable tool for helping farmers cope with major crises and disasters, but rely on farmer's and other stakeholders to handle and interpret data complexity (Lioutas & Charatsari, 2021). The overarching goal of developing and implementing AI algorithms is to implement human thinking, reasoning, while making a decision. Humans need to still be able to understand and control AI to ensure its ethical application, using expert knowledge to improve its predictive power, and ultimately to realize its full societal potential. This is termed "Humans-in-the-Loop (HITL)" and combines the power of both artificial (machine/computer) and human intelligence.

Canada's digital economy was estimated in 2018 to be worth $118 billion in terms of gross domestic product (GDP), outpacing the rest of the economy (5.5% of total economic activity) of $200 billion by 2050.[22] Digital agriculture is still maturing and developing in terms of culture, governance, technology, data, and predictive analytics. What do all of these changes mean for agriculture, complex supply chains, and value chains, how farmers manage their farms, and the use of digital agriculture? At the level of entire agricultural supply chains and value chains, there is a need to build trust and transparency, internet connectivity for cloud-based automation, greater equity and opportunity for rural, developing countries, expanded technical training, and broader assessment of return on investments (ROI) for new deployed digital technologies. At the farm-scale, AI and automation still needs to be better matched to farm needs, given low levels of adoption in smaller farms. Digital solutions need to better integrate farmer's and other local expert knowledge,

[22] Statistics Canada. 2021. Digital Economy and Society Statistics. https://www.statcan.gc.ca/en/subjects-start/digital_economy_and_society

including social and natural perspectives, and indigenous knowledge. At the scientific community scale, there is a need to share data, while reducing the disaggregation and inconsistency of data, enabling data validation and accountability. For example, there are major knowledge gaps concerning the usefulness, benefits, compatibility, and interoperability of smart, digital technology shaping "smart farms" of the future (e.g., autonomous systems such as driverless tractors, fruit picking robots, and spraying drones) for different FS and their potential impact on different FS types, farm labor, and costs. There is a large divide in data-driven farming with many countries in Africa, given that less than 40% of farming households have Internet access, and the cost of data remains prohibitive. Recent, published estimates indicate that only 24–37% of farms of <1 ha in size are served by third-generation (3G) or 4G services, compared to 74–80% of farms of >200 ha in size, with many countries in Africa, more than 60% of farming households, having no internet access (Mehrabi et al., 2021). Despite many challenges, there are big opportunities for digital technology applied to diverse FS types spanning rural, peri-urban, and urban agriculture. Digital agriculture is driving the transformation and redesign of agriculture, providing an opportunity to build back better and make farms more efficient, resilient, and sustainable. This can help to ensure the knowledge and perspective of all resource stakeholders, including indigenous peoples, are represented in agricultural decision-making. Digital agriculture using big data, predictive analytics, and Internet-of-Things (IoT) can help to manage agroecosystem resources more efficiently and effectively, to enhance the environmental performance and increase the resiliency and sustainability of agricultural FS. This, however, relies on agricultural stakeholders (i.e., end-users) to improve, adopt, and integrate digital solutions. A digital inclusion agenda needs to be enhanced and expanded whereby governments, the development community, and the private sector focus their efforts to improve access so that data-driven agriculture is available to all farmers globally.

Flexible Policy Portfolios for Rapidly Changing Food Systems and Digital Solutions

The Food Policy for Canada, launched in 2019, is building a foundation for increased integration and coordination of food-related policies and programs with the vision that "All people in Canada are able to access a sufficient amount of safe, nutritious, and culturally diverse food."[23] The 2030 Agenda for Sustainable Development, adopted by Canada and all 193 United Nations member states in 2015, is a global framework centered around an ambitious set of 17 Sustainable Development Goals (SDGs), covering the systemic, interconnected economic, social, and environmental dimensions of sustainable development that calls for "transformative shifts,

[23] AAFC. 2019. Food Policy for Canada. https://agriculture.canada.ca/sites/default/files/legacy/pack/pdf/fpc_20190614-en.pdf

integrated approaches and solutions to structural barriers to sustainable development and it recognizes the fundamental role played by sustainable agriculture in the connection between people, planet and prosperity." The FAO in 2021 also convened a Food Systems Summit[24] to raise awareness and elevate public discussion about how to reform FS and achieve significant progress toward the 2030 Agenda for Sustainable Development (SDA). FAO's new Strategic Framework 2022–31[25] seeks to support the SDA through the transformation to more efficient, inclusive, resilient, and sustainable FS for better production, better nutrition, a better environment, and a better life, leaving no one behind. In transforming and redesigning urban and peri-urban FS, the Better Life program in this strategic framework supports national and local decision-makers to initiate, coordinate, and scale up actions and investments.

Achieving emission reductions by 2030 and net-zero by 2050 will redesign and transform agricultural production systems as well as changing crops grown and diets, but won't necessarily address food insecurity, unless gaps, inefficiencies in growing and distributing food sustainably and cost-effectively can also be achieved. Canada launched the Pan-Canadian Framework on Clean Growth and Climate Change in 2016 with its provinces, territories, and Indigenous peoples to reduce carbon pollution, help Canada adapt and become more resilient to the impacts of a changing climate, spur clean technology solutions, and create good jobs that contribute to a stronger economy.[26] Within this plan, the Agricultural Climate Solutions (ACS) initiative is establishing nationwide network of Living Laboratories to bring together scientists and stakeholders to develop, test, and monitor best management practices to reduce agriculture's environmental footprint, while enhancing its resiliency to climate change. It also includes an On-farm Action Fund that supports farmers in adopting beneficial management practices (BMPs) that store carbon and reduce greenhouse gases.[27] Global total agricultural GHG emissions for 2010 (2007–2013) were $17{,}318 \pm 1675\,TgCO_{2,\,eq}\,yr.^{-1}$, of which 57% corresponds to the production of animal-based food (including livestock feed), 29% to plant-based foods, and 14% to other utilizations. Farmland management and land-use change represented major shares of total emissions (38% and 29%, respectively) (Xu et al., 2021). Based on 2015 data, total agricultural GHG emissions increased to 18 Gt CO_2 equivalent per year globally, representing 34% of total global GHG emissions. The largest contribution came from agriculture and land use/land-use change activities (71%), with the remaining from supply chain activities: retail, transport, consumption, fuel production, waste management, industrial processes, and packaging

---

[24] FAO, 2021b. Food Systems Summit 2021. https://www.un.org/en/food-systems-summit/about

[25] FAO, 2021c. FAO's Strategic Framework 2022–31. Food and Agriculture Organization of the United Nations Rome, October 2021 https://www.fao.org/3/cb7099en/cb7099en.pdf

[26] Report of the Standing Committee on Agriculture and Agri-Food. Toward a Resilient Canadian Agriculture and Agri-Food System: Adapting to Climate Change. 2018. 42nd PARLIAMENT, first SESSION, 70 pp. https://www.ourcommons.ca/Content/Committee/421/AGRI/Reports/RP9814809/agrirp11/agrirp11-e.pdf

[27] Agriculture and Agri-Food Canada (AAFC), 2021. Agricultural Climate Solutions (ACS). https://agriculture.canada.ca/en/agriculture-and-environment/agricultural-climate-solutions

(Crippa et al., 2021). In Canada, total agriculture GHG emissions (not factoring in carbon sequestration by agricultural soils) amounts to 8% of Canada's total emissions. Only Japan, Canada, and the European Union (EU) have legally binding net-zero commitments. For example, Sweden and Germany have legally binding net-zero targets for 2045, while France, Denmark, Spain, Hungary and Luxemburg have set theirs for 2050. Japan, Korea, Canada, and New Zealand have all passed laws committing to achieving net-zero by 2050 while there is proposed legislation in Ireland, Chile, and Fiji.[28] Most countries are lagging behind, with only two countries having achieved carbon neutrality, namely, Suriname in 2009 and Bhutan in 2014.

Flexible policy responses to address food security and nutrition are urgently needed, especially given the impacts of COVID. Food production and trade is a global business and national policies need to take into account such trade, rather than exporting problems to other countries. Comprehensive portfolios of policies, investments, and legislation are needed to maximize the synergistic effects for achieving food system transformation, while mitigating undesirable trade-offs. Potential mechanisms that could mitigate the impacts of major crises or disasters in agriculture are resilience-promoting policies, community marketing schemes, and smart farming technologies (Lioutas & Charatsari, 2021). Current FS, which are undergoing transformation and redesign, will also, in turn, reconfigure regional food-supply networks. Local to regional changes to improve resiliency will need to be interlinked in decision-making and intervention actions. To increase food self-sufficiency, urban areas, in particular, will need to minimize their dependence on far away resources and provisions by involving the selection of alternative, regional supply zones both in times of stability and emergency. COVID, along with extreme events, continues to impact communities, infrastructure, supply-chains, and livelihoods around the world. COVID has demonstrated how vital sound, evidence-based science, transparency in decision-making, and communication under uncertainty are to ensuring resiliency as wave after wave has spread. It has also highlighted how rapid fake news, misinformation, confusion, and miscommunication itself can propagate and spread.

Citizens must be placed front and center when redesigning and transforming FS, learning from success stories and demonstration use-cases globally. Redesign and transformation of FS is needed, not just to build back greater resiliency in current systems. Citizens themselves have a collective role and responsibility to build resiliency in FS, not just rely and react to responses and decisions of governments, private companies, nongovernmental organizations, and industrial sector commodity organizations. Citizen science solutions, proactive actions by sponsored community groups, and crowdsourcing services can help combat misinformation, democratize science encouraging codesign. Bottom-up campaigns and community initiatives,

[28] Carver, D. 2021. Insight. Global net zero commitments. United Kingdom (UK) Parliament. House of Commons Library. Published online: 12 November, 2021. https://commonslibrary.parliament.uk/global-net-zero-commitments/#:~:text=Sweden%20and%20Germany%20have%20legally,and%20Fiji%20have%20proposed%20legislation

not just top-down guidelines, protocols, and incentives, are needed to fully promote and encourage the adoption of digital technologies and practices for better anticipating, assessing risks, and building resiliency in FS. What is clear is the urgency for us to learn, adapt, and prepare to a wide array of possible unanticipated shocks, emergencies, and disasters under climate change.

Concluding Remarks

Food insecurity is complex, with countries interdependent on food imports and exports to varying degrees, with their situations changing rapidly during times of emergencies and disasters. Water is a critical input for agricultural production and plays an important role in food security, so managing water sustainably is key to the future of agriculture and achieving climate change and global SDGs. NbS offer an integrated approach to enhancing the resilience and sustainability of FS, but requires strong engagement and coordination of stakeholders to codesign, pilot, and evaluate strategies and actionable solutions to be effective. Potential mechanisms that could mitigate the impacts of major crises or disasters in agriculture are resilience-promoting policies, community marketing schemes, and smart farming technology. Citizens must be placed front and center when redesigning and transformation FS, learning from success stories and demonstration use-cases globally. Redesign and transformation of FS is needed, not just build back greater resiliency in current systems. Citizens themselves have a collective role and responsibility to build resiliency in FS, not just rely and react to responses and decisions of governments, private companies, nongovernmental organizations, and industrial sector commodity organizations. Citizen science solutions, proactive actions by sponsored community groups, and crowdsourcing services can help combat misinformation, democratize science encouraging codesign. Bottom-up campaigns and community initiatives, not just top-down guidelines, protocols, and incentives, are needed to fully promote and encourage the adoption of digital technologies and practices for better anticipating, assessing risks, and building resiliency in FS. What is clear is the urgency for us to learn, adapt, and prepare to multiple, unanticipated shocks, emergencies, and disasters under climate change.

References

Ayed, R. B., & Hanana, M. (2021). Artificial intelligence to improve the food and agriculture sector. *Journal of Food Quality, 5584754*, 7.

Boero, V., Cafiero, C., Gheri, F., Kepple, A. W., Rosero Moncayo, J., & Viviani, S. (2020). *Access to food in 2020. Results of twenty national surveys using the food insecurity experience scale (FIES)*. FAO.

Bojić, D., Baas, S., & Wolf, J. (2019). *Governance challenges for disaster risk reduction and climate change adaptation convergence in agriculture. Guidance for analysis*. Governance and Policy Support. Discussion paper. FAO.

Carisse, O., Fall, M., & Vincent, C. (2017). Using a biovigilance approach for pest and disease management in Quebec vineyards. *Canadian Journal of Plant Pathology., 39*, 393–404.

Challinor, A., & Benton, T. G. (2021). International dimensions. In R. A. Betts, A. B. Haward, & K. V. Pearson (Eds.), *The third UK climate change risk assessment technical report*. Prepared for the Climate Change Committee.

Crippa, M., Solazzo, E., Guizzardi, D., et al. (2021). Food systems are responsible for a third of global anthropogenic GHG emissions. *Nature Food, 2*, 198–209.

Deaton, B. J., & Deaton, B. J. (2020). Food security and Canada's agricultural system challenged by COVID-19. *Canadian Journal of Agricultural Economics, 68*(2), 143–149.

Deaton, B. J., & Deaton, B. J. (2021). Food security and Canada's agricultural system challenged by COVID-19: One year later. *Canadian Journal of Agricultural Economics, 69*, 161–166.

Fanzo, J., Haddad, L., McLaren, R., et al. (2020). The food systems dashboard is a new tool to inform better food policy. *Nature Food, 1*, 243–246.

First Nations Information Governance Centre. (2018, March). National Report of the First Nations Regional Health Survey Phase 3: Volume One (181 pp).

HLPE. (2017). Nutrition and food systems. A report by the High Level Panel of Experts on Food Security and Nutrition of the Committee on World Food Security, Rome. Available online: https://openknowledge.fao.org/handle/20.500.14283/i7846e

Holleman, C., Rembold, F., Crespo, O., & Conti, V. (2020). The impact of climate variability and extremes on agriculture and food security—An analysis of the evidence and case studies. In *Background paper for the state of food security and nutrition in the world 2018*. FAO.

Klerkx L, Begemann S. (2020). Supporting food systems transformation: The what, why, who, where and how of missionoriented agricultural innovation systems. Agric Syst. 2020 Sep;184:102901. https://doi.org/10.1016/j.agsy.2020.102901

Lioutas, E. D., & Charatsari, C. (2021). Enhancing the ability of agriculture to cope with major crises or disasters: What the experience of COVID-19 teaches us. *Agricultural Systems, 187*, 103023.

Marusak, A., Sadeghiamirshahidi, N., Krejci, C. C., Mittal, A., Beckwith, S., Cantu, J., Morris, M., & Grimm, J. (2021). Resilient regional food supply chains and rethinking the way forward: Key takeaways from the COVID-19 pandemic. *Agricultural Systems, 190*, 103101.

Mehrabi, Z., McDowell, M. J., Ricciardi, V., Levers, C., Martinez, J. D., Mehrabi, N., Wittman, H., Ramunkutty, N., & Jarvis, A. (2021). The global divide in data-driven farming. *Nature Sustainability, 4*, 154–160.

Moberg, E., Allison, E. H., Harl, H. K. et al. (2021). Combined innovations in public policy, the private sector and culture can drive sustainability transitions in food systems. *Nat Food 2*, 282–290. https://doi.org/10.1038/s43016-021-00261-5

Newlands, N. K. (2016). Future sustainable ecosystems: Complexity, risk, uncertainty. In *Taylor and Francis Group LCC, (Chapman & Hall/CRC applied environmental statistics series), Boca Raton, Florida, USA*. ISBN 978-1466582569 (p. 409). CRCPress.

Newlands, N. K. (2021). Big data governance, technology, and implementation in climate-resilient societies. In R. C. Brears (Ed.), *The Palgrave handbook of climate resilient societies*. Palgrave Macmillan.

Pretty, J. (2020). New opportunities for the redesign of agricultural and food systems. *Agriculture and Human Values, 37*, 629–630.

Pretty, J., Benton, T. G., Bharucha, Z. P., Dicks, L. V., Flora, C. B., Godfray, H. C. J., Goulson, D., Hartley, S., Lampkin, N., Morris, C., Pierzynski, G., Prasad, P. V. V., Reganold, J., Rockström, J., Smith, P., Thorne, P., & Wratten, S. (2018). Global assessment of agricultural system redesign for sustainable intensification. *Nature Sustainability, 1*, 441–446.

Savary, S., Akter, S., Almekinders, C., et al. (2020). Mapping disruption and resilience mechanisms in food systems. *Food Security, 12*, 695–717. https://doi.org/10.1007/s12571-020-01093-0

Sonneveld, B. G. J. S., Merbis, M. D., Alfarra, A., Ünver, O., & Arnal, M. A. (2018). Nature-based solutions for agricultural water management and food security. In *FAO land and water discussion paper no. 12* (66). FAO.

Tubiello, F. N., Wanner, N., Asprooth, L., Muller, M., Ignaciuk, A., Khan, A. A., & Rosero Moncayo, J. (2021). Measuring progress towards sustainable agriculture. In *FAO statistics working paper* (pp. 21–24).

Vanbergen, A. J., Aizen, M. A., Cordeau, S., Garibaldi, L. A., Garratte, M. P. D., Kovács-Hostyánszkif, A., Lecuyer, L., Ngog, H. T., Potts, S. G., Settele, J., Skrimizea, E., & Young, J. C. (2020). Transformation of agricultural landscapes in the Anthropocene: Nature's contributions to people, agriculture and food security. *Advances in Ecological Research, 63*, 193–253.

Weersink, et al. (2021). COVID-19 and the Agri-food system in the United States and Canada. *Agricultural Systems, 188*, 103039.

Willett, W., Rockström, J., Loken, B., Springmann, M., Lang, T., Vermeulen, S., Garnett, T., Tilman, D., DeClerck, F., Wood, A., & Jonell, M. (2019). Food in the Anthropocene: The EAT–lancet commission on healthy diets from sustainable food systems. *The Lancet, 393*(10170), 447–492.

Xu, Sharma, P., Shu, S., et al. (2021). Global greenhouse gas emissions from animal-based foods are twice those of plant-based foods. *Nature Food, 2*, 724–732.

Chapter 3
Environmental Resilience and Agricultural Best Practices

Robert Smith

Abstract Lake Erie is increasingly stressed by hazardous and nuisance algal blooms (HNABs) that occur each summer due to excess phosphorous (P) loadings from urban and, especially, agricultural sources. Building on an earlier economic analysis of the direct impacts of these HNABs, this study estimates the full economic impacts on the Ontario economy using a regional economic impact model. The full economic impacts include indirect and induced impacts in addition to the direct impacts associated with the costs previously reported in the literature. Province-wide, 3024 person years of employment are predicted to be lost on average annually in the period 2020–2024 if nothing is done to control phosphorus loadings to the lake. Employment losses are reduced by 812 person years (to 2212) if policy action is taken to control loadings. The equivalent figures for GDP are $147,351,000 and $38,431,000, respectively.

Keywords Algal blooms · Lake Erie · Resilience · Urban agriculture

Introduction and Background to the Study

Since the mid-1990s, a resurgence of planktonic cyanobacterial harmful algae blooms has been noted in Lake Erie. Increased loadings of phosphorus in a highly reactive form (dissolved reactive phosphorus) are thought to be in large part responsible for the return of these blooms, not seen since the 1970s. Increased summertime temperatures are also a factor. With the expected rise in global temperature, the likelihood of occurrence of algae blooms in water bodies around the world is significantly increasing. This poses serious problems for the production of food particularly near urban centers close to water.

R. Smith (✉)
Midsummer Analytics, Ottawa, ON, Canada
e-mail: rob@midsummer.ca

A. A. Kubursi et al. (eds.), *Urban Food Security in a Crisis Prone World*,
https://doi.org/10.1007/978-3-031-89440-4_3

Free-floating blooms of *Microcystis aeruginosa* and other toxin-producing cyanobacteria[1] form annually in the water column of the lake's shallow, warm western basin and along the shorelines of the deeper, cooler central and western basins. Blooms are developing earlier and remaining present later in the year as time goes by.

If people drink water contaminated by microcystins (the toxin produced by *Microcystis aeruginosa*), symptoms of exposure include nausea, vomiting, and, in rare cases, acute liver failure. Minor skin irritation can also occur with contact. Reported health effects from cyanobacteria in humans are uncommon in the United States and Canada. The effects of toxic algae on fish are mostly due to the hypoxia induced by decaying algae rather than from the algae itself.

Some areas of Lake Erie's western basin are also affected by dense bottom-resting mats of *Lyngbya*, which is a nontoxic but odorous cyanobacteria. In the eastern basin, large shoreline blooms of the bottom-attached filamentous green algae *Cladophora*[2] are common. These blooms foul recreational beaches with dense covers of dead algal "muck," clog municipal and industrial water intakes, impair water quality, and can also pose microbial health risks[3] to wildlife and humans.

As with hazardous blooms, increased loadings of dissolved reactive phosphorus—mainly from run-off of fertilizer from farmland—are a major cause of the growth of nearshore nuisance blooms. Other factors are increased summertime temperatures and, in the case of *Cladophora*, the presence of dreissenid (zebra and quagga) mussels, which filter water (allowing greater light penetration at depths where the algae form) and provide a substrate for algal growth.[4]

To reduce the severity of HNABs, the governments of Canada and the United States have committed to reducing phosphorus loads to Lake Erie by 40% from 2008 levels and coordinate action through a binational framework developed under the Great Lakes Water Quality Agreement (Environment and Climate Change Canada and Ontario Ministry of the Environment and Climate Change, 2018). Work is underway to assess whether a similar phosphorus reduction target is required to meet various water quality and ecosystem health improvements for Lake Ontario. A significant element in the plan seeking to achieve the required phosphorous reductions relates to the implementation of best management practices (BMPs) on farmland. Environment and Climate Change Canada, the federal ministry of the environment, has carried out numerous studies in recent years to understand the potential for phosphorous reductions from agricultural BMPs and the economics of

[1] Cyanobacteria, or "blue-green algae" are not, strictly speaking, algae but, in fact, large masses of bacteria.

[2] *Cladophora* is a true form of algae.

[3] Though nuisance algae do not produce toxins, they can harbor harmful bacteria (such as *Escherichia coli*).

[4] The material in the preceding paragraphs is drawn from International Joint Commission, 2013 and Michigan Sea Grant (n.d.).

their implementation. Smith et al. (2019)[5] describe the results of one of these studies. Using a welfare economics framework, Smith et al. considered the economic costs of HNABs on a variety of users of Lake Erie's ecological goods and services resident in the Canadian portion of the lake's watershed basin:

- Commercial fishers.
- Industrial and municipal water users.
- Tourism operators.
- Property owners.
- Recreational users.
- Nonusers (that is, those who care about the lake's quality for intrinsic reasons).

Smith et al. estimated that HNABs, if left unchecked, would impose total costs (utility losses) on these users equal to $5324 million (2015 dollars) over a 30-year period. If action was taken to control loadings of phosphorus to the lake, such that total loadings from the Canadian side of the basin declined by 40%, 30-year costs were found to decline to $2474 billion.

Smith et al. considered only utility losses imposed directly on the lake's users. They did not consider the broader impacts of HNABs on the regional economy. Such costs may arise because those who suffer utility losses due to HNABs are apt to alter their spending behavior in ways that will have impacts, both positive and negative, on the broader economy. Fishers who lose income due to reduced catches will spend less on market goods and services. Industrial and municipal users that must adjust their operations to account for the presence of algal biomass and toxins in their source water will face higher equipment and operating costs as a result. Tourism operators who lose business because of polluted water and beaches will employ fewer staff and earn less market income. Property owners who lose equity in their homes will reduce their market spending to offset their losses. The pure utility losses that recreational and nonusers face may be thought of in terms of equivalent losses in market spending. The purpose of the present study was to build upon Smith et al.'s work by estimating the broader economic impacts (GDP, employment, tax collection, etc.) of the changes in spending behavior induced by HNABs.

Adopting a standard economic impact analysis approach, this study considered both the direct impacts of the changes in economic activity induced by HNABS and their secondary (indirect and induced) impacts.

Costs in Smith et al. were considered under two scenarios: "business-as-usual," in which nothing is done to limit phosphorous (P) loading and the lake's condition

[5] The author of the present report was also lead author for Smith et al. (2019). Smith et al. was based on the results of a 2015 study conducted by Midsummer Analytics of Ottawa, Ontario (of which the author is the principal) in collaboration and EnviroEconomics of Ottawa, Ontario for Environment and Climate Change Canada. The present study is based on the results of a subsequent (2019) study conducted by Midsummer Analytics in collaboration with Econometric Research Ltd. of Burlington, Ontario and EnviroEconomics, again for Environment and Climate Change Canada. The results of the 2019 study are reproduced here with the permission of the study's technical authority, Dr. Brad Bass, Great Lakes Harmful Pollutants Branch, Environment and Climate Change Canada.

steadily worsens; and "policy intervention," in which P loadings to the lake are reduced through government action assumed to begin in 2020 and HNAB events on the lake decrease in severity and frequency. The approach and assumptions used in Smith et al. were intentionally conservative so as not to overstate the costs of HNABs. Nonetheless, the costs were, as noted, found to be significant under both scenarios, but much more so (more than double) if nothing was done to control algal blooms (business-as-usual).

Smith et al. considered only the costs of HNABs imposed on those who consume the lake's EGS directly, such as commercial fishers, beach goers, and municipal drinking water plants. While relevant to understanding the economic consequences of HNABs, these costs do not represent the full economic impact of the blooms. The full impacts include (1) the impacts on the businesses and workers who provide the direct users of the lake's EGS with market products and labor ("direct" impacts in impact modeling parlance); (2) the impacts on other sectors of the economy that do business with the directly impacted sectors (indirect impacts); and (3) the impacts on the economy due to changes in household consumption as a result of direct and indirect impacts (induced impacts). The purpose of this study was to estimate the direct, indirect, and induced economic impacts of the direct costs imposed on users of the lake's ecological good and services. Naturally, the direct impacts are significantly smaller than the total impact, which is the sum of the indirect, induced, and direct impacts.

Multiregional Input–Output Analysis and Economic Impact Modeling

To convert the costs imposed on users of Lake Erie's EGS due to HNABs from Smith et al. into estimates of their broader economic impacts, a multiregional input–output (MRIO) model was required. For this purpose, we used a proprietary MRIO developed by Econometric Research Ltd. (ERL) of Burlington, Ontario. More specifically, the MRIO used was a special version of the open and generic Regional Economic Impact Model for Canada (REIM-Canada) developed by ERL. REIM-Canada operates at the level of counties, provinces and Canada as a whole. Here, the Ontario regional sub-model of REIM-Canada (REIM-Ontario) has been used.

REIM-Canada is a specific implementation of the generalized MRIO first proposed by Isard (1960). Isard showed that gross output by industry at the regional scale could be expressed as a function of intra-regional technical coefficients and interregional trade coefficients representing the economic structures of all industries in each region and the nature of the interconnections between industries across regional boundaries.

REIM-Canada is based on an integration of (1) input–output analysis; (2) location theory (see Box 3.1); and (3) relevant segments of typical macroeconomic models. The model utilizes a large set of economic and technical databases that are regularly published by Statistics Canada (Canada's national statistical agency). A short list includes the inter-provincial input–output tables; employment by sector; taxes by type and level of government; product prices; and energy consumption.

REIM-Ontario estimates economic impacts on 36 economic sectors in Ontario (see 0) by county/region, interregionally (that is, between counties/regions) and for the province as a whole. ERL has used the model extensively in past studies of the economic impacts of agriculture, tourism, Great Lakes remedial action plans, climate change, and many others in Ontario.

Economic impacts may be estimated using REIM-Canada in terms of the following variables:

- **GDP (value added):** The value added for each sector of the economy (estimated as the sum of wages, rent, interest and profits, indirect business taxes, and depreciation less subsidies).
- **Gross output:** The value of sales of a sector.
- **Wages and salaries:** The value of wages paid in a sector.
- **Employment:** Number of workers measured in person years (full-time equivalent jobs).
- **Taxes by level of government:** Levels of income taxes, GST, liquor and tobacco taxes, etc., collected by the federal, provincial, and municipal governments.
- **Imports:** The value of goods and services acquired from outside the province.

Given an estimated change in spending on market goods and services—a "change in initial expenditure"—REIM-Canada will estimate impacts of three types for each of the above variables:

- ***Direct impacts***—Those associated with the sectors that produce the goods and services for which spending has changed; for example, the loss in value added in a regional fishing industry due to a reduction in spending on fish products.
- ***Indirect impacts***—Those associated with sectors that provide goods and services to directly impacted sectors. The loss of a dollar of revenue for commercial fishers, for instance, results in fewer purchases of fuel, fishing gear, labor services, insurance, boat maintenance, and so on, meaning less revenue for the sectors that supply these goods and services. Those sectors, in turn, will reduce their spending on other goods and services, with corresponding losses further up the supply chain.
- ***Induced impacts***—Those that occur not because of supply-chain effects but because the workers in directly and indirectly impacted sectors have less income to spend on consumer goods and services, reducing the overall level of economic activity in the region.

The example of drinking a beer can be used to illustrate the three types of impacts. Purchasing the beer has a **direct impact** on the beer brewing industry. **Indirect impacts** associated with the beer are the barley, water, and electricity required for brewing the beer, plus the fertilizer and tractors needed to grow the barley, the steel needed for the tractor, and so on. Through the use of matrix algebra, all indirect effects relating to market goods and services are captured in the model. **Induced impacts** associated with the beer production and serving arise because the person who serves the beer earns money doing so and spends that money on other goods and services (including beer—meaning that beer produces beer!). Again, all induced impacts are captured in the model.

In the present study, the costs imposed on users of Lake Erie's EGS were introduced into REIM-Ontario as the changes in initial expenditure to estimate the broader economic impacts (direct, indirect, and induced) of HNABs on the Lake Erie Basin economy. Before this could be done, however, a number of modifications to Smith et al.'s results were required to make them suitable as inputs into REIM-Ontario. These modifications, along with other aspects of the study approach, are described next.

Box 3.1 The Importance of Location in Regional Impact Modeling

While it is reasonable to assume that the technologies used to produce a given good or service vary little across the regions of Ontario (that is, steel is made much the same way whether produced in the city of Hamilton, Ontario, or in the city of Sault Ste. Marie, Ontario), the same cannot be said for the share of markets captured by producers in different regions. It is more likely that demand for steel from southern Ontario consumers will be met from Hamilton than from Sault Ste. Marie (assuming the steel is comparable from both mills and there is little product differentiation). Similarly, demand from northern consumers is more likely to be met from Sault Ste. Marie. For this reason, REIM-Canada uses different market share assumptions for each commodity and industry in each region of the province and these shares change with the configuration of the spatial interactions. Each region is allotted a unique trade matrix depending on its constellation of sectors (that is, how many regional, provincial, and national industrial sectors the region's industrial base comprises and their capacities) and on the distances between the region and other regions of the province. This integration of distance is crucial to the analytical power of REIM-Canada. More typical interregional models assume that market demands are simply met by regions in proportion to their share of provincial employment in a given sector. To carry on with the steel example, if the total provincial employment in steel manufacturing is 1000 and 900 of these workers in Hamilton, a typical model would allocate 90% of steel demand to Hamilton regardless where in the province the demand originated.

The potential of REIM-Canada is amplified considerably when one considers the task of measuring the economic impact of activities in one region the provincial economy on all others. This requires detailed knowledge of how the individual sectors in the various regions relate to one another on both the industrial and spatial levels. In other words, data on how production by a firm in one industry requires inputs from firms in other industries across all regions of the multiregional system. In REIM-Canada, general trade coefficients are modified to reflect the role of distance. Each region where impacts are initiated becomes the "centre of the universe" and all interactions are determined in relation to how near or far are they from it. If ten regions are to be modeled, 100 assignment (trade) matrices are required in the model, one for each possible combination of two regions with one the importing and the other the exporting region.

Approach

Determining the Regional Scope of Analysis

REIM-Ontario is designed to produce estimates of economic impacts at the level of Ontario counties and regions. As such, it requires changes in initial expenditures to be estimated at the same level. Smith et al., in contrast, presented their cost estimates at the level of the "Lake Erie basin economy." Since this "economy" does not exist in any strict sense of the word, Smith et al. defined it variously to include household and business activities taking place within 1, 50 and 100 kilometers of the lake's shoreline depending on the cost category in question.

In order to match the geographic boundaries used by Smith et al. as closely as possible, this study focused on the following ten counties/regions in southern Ontario: Niagara Region, Haldimand County, Norfolk Country, Elgin County, Chatham-Kent County, Essex County, Brant County, Oxford County, Middlesex County, and Waterloo Region. Depending on the cost category in question, impacts were estimated in this study for all these counties/regions or for different sub-sets of them.

Selecting the Analytical Period

Though Smith et al. considered costs from 2015 to 2045, the present study considered impacts only for the period 2015 to 2024 to assure the reliability of the results. Estimating impacts beyond 2024 would have implied that MRIO models, like REIM-Ontario, produce robust estimates well into the future. Though every effort was made to ensure that REIM-Ontario could be used reliably to project impacts over the 10-year period considered in the study,[6] the dynamics of the economy are such that prediction beyond this 10-year period at the level of detail generated by MRIO models was not considered viable.

To reduce the complexity of the analysis in terms of model inputs and results, rather than preparing an annual time series of changes in initial expenditure by county/region, it was decided to prepare average changes in initial expenditure for two periods: 2015–2019 and 2020–2024. Producing an annual time series of expenditure changes would have added dozens of additional tables of results for limited gain in analytical understanding.

[6] For example, historical (2005–2015) rates of change in the key model parameters (such as rates of change in market share by commodity and industry) were built into the model to take account of the medium-term dynamics of the economy.

Recasting Costs by Geographic Region

The results from Smith et al. could not be used "as-is" because the Lake Erie basin economy was, as discussed earlier, defined geographically using a novel approach in Smith et al. Adjustments were therefore needed to apportion Smith et al.'s results to the geographic boundaries used in REIM-Ontario (counties/regions of Ontario). The approach to this apportioning, which differed from one cost category to another, is outlined in Table 3.1.

Table 3.1 Approach to apportioning basin-wide costs from Smith et al. to regions/counties, by cost category

Cost category and costing method in Smith et al.	Apportioning approach	Comments
Commercial fishing—Smith et al.'s estimated costs as losses in value added by the commercial fishing fleet on Lake Erie due to increased fishing effort and/or decreased catch of commercial fish species	Basin-wide costs from smith et al. were apportioned to counties/ regions in proportion to the share of basin economy-wide employment in fishing, hunting, and trapping in each of the six counties/regions that abut the lake. Employment data were taken from statistics Canada's census of population	Assumptions: • Commercial fishing only occurs in counties/regions that abut the lake • The share of hunting and trapping employment is small enough to ignore
Water users—Smith et al. estimated costs as increased capital and operating costs imposed on drinking water treatment plants and industrial water consumers due to the increased effort required to treat intake water from the lake before use/distribution	Basin-wide costs for municipal drinking water plants from smith et al. were allocated to counties/ regions on the basis of the location of the facility	The costs in smith et al. for industrial facilities were almost negligible (30-year NPVs of less than one million) and were ignored in the present study
Commercial tourism—Smith et al. estimated costs as losses in value added for the commercial tourism industry due to reduced tourist flows to the region resulting from the lake's reduced attractiveness as a vacation site	Basin-wide costs from smith et al. were apportioned to counties/ regions (except Waterloo region) based on tourism employment	Waterloo region was excluded from the analysis because its tourist activity was deemed to be driven by the city itself and not by Lake Erie Tourism employment was calculated as the sum of employment in the following industries in the north American industry classification system (NAICS): Arts, entertainment, and recreation (NAICS 71); accommodation and food services (NAICS 72); and transportation and warehousing (NAICS 48–49). It is recognized that not all employees in these industries are engaged in tourism activities, so there is some degree of inaccuracy in the approach

Table 3.1 (continued)

Cost category and costing method in Smith et al.	Apportioning approach	Comments
Property values—Smith et al. estimated costs as losses in the value of residential properties located along (<1 km) the lake's shoreline due to the impact of blooms on neighborhood desirability and, therefore, home prices	The costing model for property value losses from smith et al. was rerun with data on average residential property prices and number of properties by county/region for the six counties/regions that abut the lakeshore	
Recreational users—Smith et al. estimated costs as losses in utility suffered by Canadian beach goers, boaters, fishers, hunters, and wildlife viewers within 50 km of the Lake Erie shoreline	The Johnston and Thomassin (2010) regression model used to estimate willingness to pay of recreationists for freshwater quality improvements in Smith et al. was rerun using population and average household income data (from Statistics Canada) for all ten counties/regions. The revised willingness-to-pay values by county/region were then rerun through the costing model for recreationists from Smith et al.	The population and average household incomes of the ten counties/regions used in this study do not match exactly the population and average incomes of households living within the 50 km boundary that was used to define the basin economy for nonusers in smith et al.
Nonusers—Smith et al. estimated costs as losses in utility suffered by Canadians living within 100 km of the Lake Erie shoreline who do not directly use Lake Erie's ecological goods or services but who value its quality nonetheless	The Johnston and Thomassin (2010) model used to estimate willingness to pay of nonusers for freshwater quality improvements in Smith et al. was rerun using population and average household income data (from Statistics Canada) for all ten counties/regions. The revised willingness-to-pay values by county/region were then rerun through the costing model for nonusers from Smith et al.	The population and average household incomes of the ten counties/regions used in this study do not match exactly the population and average incomes of households living within the 100 km boundary that was used to define the basin economy for nonusers in smith et al.

Estimating Market Expenditure Equivalents of Losses in Property Value and Pure Utility Losses for Recreationists and Nonusers

Ordinarily, economic impact analysis considers impacts only in terms of changes in expenditure on production and consumption of market goods and services. Anything related to nonmarket goods and services is generally considered out of scope. The scope of the study carried out by Smith et al. was, however, broader than this. They considered the impacts of HNABs on consumption of Lake Erie's *nonmarket* EGS; specifically, impacts on the enjoyment of the lake's recreation services (swimming, boating, birdwatching, and hunting) and on the enjoyment that nonusers get from simply knowing of the lake's existence.

In addition, Smith et al. considered disinvestment in produced capital stocks, something also not ordinarily dealt with in impact analysis; more specifically, Smith et al. considered the loss of residential property values for homeowners along the lake's shoreline.

In neither of the above cases did the results from Smith et al. provide the changes in initial spending required as inputs for REIM-Ontario. Their results for recreational and nonusers were in the form of pure nonmarket utility losses. Their results for property owners were in the form of changes in the value of residential properties. As a result, the costs imposed on these user categories had to be converted into equivalent changes in market spending.

Converting Losses in Property Values to Declines in Market Expenditure

The relationship between wealth and expenditure has been a topic of interest in economics at least since to the work of Keynes in the 1930s. Wealth—or the value of the assets possessed by an individual—comes in a variety of forms: the wealth associated with the individual's own skills and capacities, which she/he uses to generate employment income; the wealth associated with financial assets (such as corporate shares) owned by the individual; and the wealth tied up in the physical assets owned by the individual (most importantly, their primary residence). Since wealth represents the value of future income, individuals who hold relatively more wealth are likely to consume more of their income (as opposed to saving it) compared with less wealthy individuals, as they understand they can count on their wealth as a "cushion" if their income were ever to drop.

Human wealth (or human capital, as it is better known) is the most important source of income (wages) for almost all individuals and is, therefore, a strong determinant of most individuals' potential for market expenditure. Financial assets and physical assets are, generally speaking, less important as sources of income to most people, but they remain significant since they influence savings; wealthy households do not need to spend the same proportion of their incomes as poorer households. Physical assets, especially homes, are found to play a more important role in driving market expenditure than financial assets, likely because people view changes in financial wealth as transient (stock markets swing up and down) and fewer people own financial assets than do homes (Pichette, 2004; Winkler, 2016; Coskun et al., 2018). Individuals whose houses continually rise in value, on the other hand, will be likely to spend a greater share of their income on the grounds that housing markets are generally stable and, therefore, they will one day be able to cash in the equity accumulated in their houses.

The relationship between changes in housing wealth and market expenditure has been the subject of considerable study. The impact of a change in housing wealth on expenditure is generally found to be around 5%; in other words, for a $1 increase in the value of an individual's home, his/her market expenditure will likely go up by about 5 cents. In a study published by the Bank of Canada, the figure was found to be 5.7% for Canada (Pichette, 2004). This figure has been applied here to convert losses in residential property values due to HNABs to corresponding losses in market expenditure.

Converting Utility Losses to Declines in Market Expenditure

Recreationists and the so-called "nonusers" (individuals who derive benefits from Lake Erie's simple existence rather than from any active use of it) bore important costs in the Smith et al. study. These users are affected by declines in the lake's quality because they receive less enjoyment from their recreational experiences or, in the case of nonusers, less enjoyment from their knowledge of the lake's ecological condition.

A central concept in welfare economics is the notion that there exists an amount of market income an individual could be given that would exactly offsets his/her loss in utility when a nonmarket good is lost or degraded; for example, the loss in utility from recreational enjoyment when Lake Erie is degraded (Winch, 1973). In theory, this amount, known as an equivalent variation, can be calculated according to Eq. 3.1 for a given nonmarket good:

$$U\left(P^0, Q^0, y^0 + E\right) = U\left(P^0, Q^1, y^0\right) \tag{3.1}$$

where:

U is the indirect utility optimized for income, prices of market goods, and the quality of the nonmarket good.
P is a vector of prices of market goods.
Q is the quality of the nonmarket good.
Y is market income.
E is the equivalent variation, and.
superscripts 0 and 1 refer, respectively, to the condition of the nonmarket good under the status quo (degraded state) and some hypothetical improved state.[7]

Though it is possible, in theory, to measure equivalent variation following Eq. 3.1, doing so in practice is difficult. It requires knowledge of the optimal level of demand for market goods (optimal demand functions derived from utility maximization) that is not readily available.

Another way to estimate equivalent variation is by measuring the difference in utility between the two states of the nonmarket good, since *E* is intended to exactly offset this loss. This approach has been used here.

Smith et al. used a model of willingness to pay for freshwater quality improvements (Johnston & Thomassin, 2010) to estimate household willingness to pay to improve Lake Erie's water quality by various amounts (e.g., from boatable to fishable to swimmable). This willingness to pay was taken by them as a proxy for the utility loss households in the Lake Erie basin economy face as a result of the degraded state of the lake. This utility loss is taken here as a reasonable estimate of *E*, or the amount of income households would need to receive given Lake Erie's

[7] Note that in the case of Eq. 3.1, neither the prices of market goods nor market income are assumed to be affected by the move of the nonmarket good from the degraded to improved state.

degraded state in order for them to be at the same utility level they would be if the lake were cleaner.

With this estimate of E in hand, all that was required to estimate the indirect and induced economic impacts of the utility losses for recreationists and nonusers was an estimate how much of E would be spent on market expenditure (versus being saved). For this, we used a figure of 95%, which reflects a household saving rate of 5%. Statistics Canada reports an average long-term household saving rate (1990–2015) of 5.1%.[8]

It is important to recognize that households do not, in reality, receive E; that is, they are not compensated for their lost utility with extra market income when Lake Erie degrades. So, in estimating the indirect and induced impacts of the portion of E that is spent, we are not measuring economic impacts that would actually occur anywhere in the economy. Rather, we are providing an estimate of how large those impacts would be if recreationists and nonusers actually lost market income (rather than nonmarket utility) as a result of HNABs. In this way, the total impacts of these nonmarket utility losses can be compared against the other (market) impacts considered by Smith et al.

Estimating Changes in Initial Expenditures by Category of Goods

The final step in adapting the costs from Smith et al. for use in REIM-Ontario was to convert the costs (suitably modified as discussed above) into spending on the set of goods and services built into REIM-Ontario. The approach varied by cost category.

For commercial fishers, the costs imposed by HNABs in terms of lost value added from Smith et al. were converted into declines in spending on goods and services by fishers across their normal pattern of spending. For municipal drinking water plants, the Smith et al. increased operating and capital costs were allocated to the relevant goods and service categories as well as investment expenditure. For tourism operators, costs in terms of lost value added were converted into declines in spending on goods and services (including labor) across their normal pattern of spending. For property owners, the market expenditure equivalent of lost property values was converted into declines in spending on goods and services across the normal pattern of household spending. The same approach was used for recreational users and nonusers.

[8] Statistics Canada, Table 36-10-0112-01. *Current and Capital Accounts—Households, Canada, Quarterly.*

Selected Results

In this section, we present a selection of the results from the impact modeling. As the scope of the analysis was broad and the model outputs are detailed, many results could be presented. We focus below on those we believe are of most interest; that is, the province-wide impacts by cost category. The results presented in Tables 3.2, 3.3, 3.4 and 3.5 summarize the average annual total province-wide economic impacts by cost category under each of the time periods and scenarios modeled. As a reminder, the cost categories considered in the modeling were:

- Lost commercial fishing value added.
- Increased capital and operating costs for drinking water treatment plants.

Table 3.2 Average annual total province-wide economic impacts by cost category Business-as-usual scenario, 2015–2019

Cost category	GDP	Gross output	Wages and salaries	Employment
	2017 constant dollars			person years
Lost commercial fishing value added	$3,922,850	$9,302,825	$2,724,346	70
Increased drinking water treatment plant costs	-$870,078	$5,054,435	$99,029	14
Lost utility for recreationists	$8,283,107	$25,809,917	$5,725,237	145
Lost utility for nonusers	$50,644,201	$157,805,928	$35,004,999	887
Lost property value	$3,666,864	$11,425,864	$2,534,517	64
Lost tourism value added	$44,239,210	$105,213,010	$29,905,127	993
Total	**$109,886,154**	**$314,611,979**	**$75,993,255**	**2172**

Source: REIM-Ontario
Note: Positive values represent economic losses and negative values gains

Table 3.3 Average annual total province-wide economic impacts by cost category Policy intervention scenario, 2015–2019

Cost category	GDP	Gross output	Wages and salaries	Employment
	2017 constant dollars			person years
Lost commercial fishing value added	$3,917,685	$9,290,572	$2,720,758	70
Increased drinking water treatment plant costs	-$869,967	$5,053,760	$99,014	14
Lost utility for recreationists	$7,771,138	$24,214,634	$5,371,375	136
Lost utility for nonusers	$50,791,138	$158,263,771	$35,106,555	889
Lost property value	$3,665,944	$11,422,974	$2,533,877	64
Lost tourism value added	$43,209,495	$102,784,541	$29,205,349	970
Total	**$108,485,433**	**$311,030,252**	**$75,036,928**	**2142**

Source: REIM-Ontario
Note: Positive values represent economic losses and negative values gains

Table 3.4 Average annual total province-wide economic impacts by cost category business-as-usual scenario, 2020–2024

Cost category	GDP	Gross output	Wages and salaries	Employment
	2017 constant dollars			person years
Lost commercial fishing value added	$3,940,613	$9,301,711	$2,740,770	70
Increased drinking water treatment plant costs	-$617,687	$3,689,517	$75,356	11
Lost utility for recreationists	$13,000,370	$40,748,862	$9,051,250	235
Lost utility for nonusers	$55,232,285	$173,122,254	$38,454,414	997
Lost property value	$8,025,462	$25,155,340	$5,587,577	145
Lost tourism value added	$67,769,500	$160,551,915	$46,031,152	1567
Total	**$147,350,543**	**$412,569,599**	**$101,940,519**	**3024**

Source: REIM-Ontario
Note: Positive values represent economic losses and negative values gains

Table 3.5 Average annual total province-wide economic impacts by cost category Policy intervention scenario, 2020–2024

Cost category	GDP	Gross output	Wages and salaries	Employment
	2017 constant dollars			person years
Lost commercial fishing value added	$3,196,539	$7,545,354	$2,223,253	57
Increased drinking water treatment plant costs	-$506,908	$3,027,811	$61,847	9
Lost utility for recreationists	$7,895,666	$24,748,467	$5,497,210	143
Lost utility for nonusers	$48,905,369	$153,290,902	$34,049,427	882
Lost property value	$3,511,245	$11,005,774	$2,444,636	64
Lost tourism value added	$45,917,872	$108,843,191	$31,177,898	1061
Total	**$108,919,783**	**$308,461,499**	**$75,454,271**	**2212**

Source: REIM-Ontario
Note: Positive values represent economic losses and negative values gains

- Lost value added for tourism operators.
- Lost property value for residential property owners.
- Lost utility for recreationists, and.
- Lost utility for nonusers.

and the scenarios modeled were:

- **Business-as-usual**, in which nothing is done to limit phosphorous (P) loading and Lake Erie's condition steadily worsens over time, and.

- **Policy intervention**, in which P loadings to the lake are reduced through government action assumed to begin in 2020 and HNAB events on the lake decrease in severity and frequency.[9]

To understand the results, it is first important to note that positive figures in the tables represent economic losses and negative figures (in red) represent gains. For example, the figure of $3,922,850 (constant 2017 dollars) in the upper left-hand cell of Table 3.2 represents the average annual total (direct, indirect, and induced) province-wide loss in GDP as a result of the drop in market consumption in the Lake Erie basin "economy"[10] due to HNAB impacts on commercial fishing value added under the business-as-usual scenario from 2015 to 2019. This figure means that, on average, in each year from 2015 to 2019, the province of Ontario as a whole will lose nearly $4 million in GDP due to lost commercial fishing value added in the Lake Erie basin economy associated with HNABs if nothing is done to control P loadings to Lake Erie. In other words, reduced spending on goods and services by commercial fishers in the Lake Erie basin economy ripples through the province as a whole to result in a total loss of nearly $4 million in GDP on average each year from 2015 to 2019.

The negative (red) values that appear in the tables for the total GDP impacts due to increased municipal drinking water treatment plant (DWTP) costs require explanation. The increased costs faced by DWPTs due to HNABs have two effects on the economy. On the one hand, they lead to increased purchases of goods and services (intermediate expenditures) by DWTPs. These increases, other things equal, increase economic activity. On the other hand, the increased costs of these purchases must be recouped by DWTPs by increasing costs to water consumers, which leads to a reduction in their intermediate and final spending on other goods and services. This lowers economic activity, other things equal. To estimate the overall impact of the increased costs faced by DWTPs, we have modeled both the increased intermediate spending of the plants and the reduced spending of water consumers. We find the net effect on province-wide GDP of these opposing forces to be small (less than $1 million) but positive (negative numbers in the tables being understood as positive economic impacts). In other words, the extra GDP generated by the increased intermediate spending of the DWTPs outweighs the loss in GDP due to reduced spending by water consumers. It should be noted that this positive impact does not extend to other variables; the impacts on gross output, wages and salaries, and employment are all negative (and also quite small). The fact that the GDP impact is positive while the others are negative is simply the result of the distribution of GDP, gross output, wages and salaries, and employment across the various industries of the economic and the differential impacts on these industries of the changes in spending due to HNABs impacts on DWTPs.

[9] Smith et al. also considered direct costs associated with continuation of the lake's 2014 condition—an ecologically implausible "stable lake" scenario—for the sake of comparison. The stable lake scenario has not been considered in this study.

[10] Recall that the Lake Erie "economy" does not exist in any strict sense but that we have defined as being the economic activity taking place in the ten counties/regions mentioned earlier.

Beyond this, several features of the total impacts by cost category are worth noting.

- First, **regardless of the variable, time period, or scenario considered, the largest impacts are associated with lost utility for nonusers and lost tourism value added.** Together, these cost categories account for about 85% of total losses in GDP across the time periods and scenarios, for example. This is consistent with the fact that all residents of the ten counties/regions are assumed to suffer utility losses due to the degradation of Lake Erie and that the tourism industry is economically important in the region and, therefore, the predicted decreases in tourism value added due to HNABs translate into large impacts relative to the other cost categories.
- Second, similar to counties/regions, **not all cost categories benefit equally from policy intervention.** Considering the period 2020–2024 (after policy intervention is assumed to take place; Tables 3.4 and 3.5), the increase in impacts in the business-as-usual scenario over the policy intervention ranges from a high of 128% for the drop in market consumption due to lost property value to a low of 13% for the drop in market consumption due to lost utility for nonusers. The differences are due to the ways in which changes in HNAB bloom intensities were assumed to affect the flows of different EGS from Lake Erie. Nonusers, for example, were assumed to suffer utility losses whenever the lake is degraded from its swimmable state, which it is much of the time up to 2024 even in the policy intervention scenario. Property values, on the other hand were not assumed to fall until the lake's quality fell below the level of *habitat suitable for some fish and wildlife* on the RFF water quality ladder. In addition to the impacts of lost property values, the impacts from lost recreational user utility are also found to be substantially reduced (64%) Vis-à-vis business-as-usual when policy intervention is taken to limit P loadings to the lake.

We discuss the results further in the next section.

Results

Only sample results will be presented below, more detailed results can be obtained from the author. The results here suggest that Lake Erie HNABs impose significant costs on the Ontario economy. If nothing was done to control P loadings (phosphorus loading) to the lake, average annual province-wide GDP losses were found to be more than $147 million (constant 2017 dollars) and average annual job losses were expected to be more than 3000 person years during the 2020–2024 period (Table 3.4). If policy intervention to control P loadings is effected beginning in 2020, these impacts fall considerably to $109 million and about 2210 person years. The fact that impacts remain significant even after policy action is initiated is a function of two factors. First, policy actions are assumed to take a number of years to have effect, so the lake's quality continues to improve beyond 2024.[11] Second, even after P loadings are reduced by 40%, the lake continues to suffer from degradation that causes utility losses, especially for nonusers who care intrinsically about its quality.

These results point to the importance of agricultural activity and its ecological consequences (freshwater eutrophication in this case) as a driver of environmental quality declines and economic losses. Though agriculture is not the only source of freshwater P loadings in the province, it is the most significant.

Of course, Lake Erie is not the only freshwater lake in Ontario to be impacted by eutrophication from excess agricultural P loads. Lake Ontario, another of the Laurentian Great Lakes, is also impacted. So too is Lake St. Clair, a relatively small lake connecting lakes Huron and Erie. Further afield, Lake of the Woods in northwestern Ontario is also impacted, as are many smaller freshwater lakes across the province. Given this, the total economic impacts of excess agricultural P loadings to freshwater in the province are plausibly twice or more the figures reported here.

Putting the Total Impacts in Context

It is useful to put the total impacts of HNABs into context. We do this here by comparing the negative employment impacts we find for HNABs with the projected job benefits for two well-known, Canadian construction mega-projects, the Gordie Howe Bridge from Windsor to Detroit and the twinning of the Trans Mountain oil pipeline from Edmonton, Alberta to Burnaby, British Columbia.

The $5.7 billion Gordie Howe International Bridge project will see a new bridge constructed between Windsor, Ontario and Detroit, Michigan, with ports of entry of both sides of the border and reconfiguration of a portion of Interstate-75 in Michigan. It is the largest ongoing infrastructure project along the Canada-US border and one of the largest ongoing projects anywhere in North America.

According to a report from Workforce WindsorEssex (2019), the 5-year construction phase of the bridge is expected to create 2500 direct and 1019 induced jobs. This equates to an average of about 700 direct and induced jobs created annually. No estimate of indirect job creation could be found. The 700 jobs expected to be created annually by construction of the new bridge is considerably less than half of the figure found here for the annual direct and induced employment losses associated with Lake Erie HNABs from 2020 to 2014 under business-as-usual (an average of 1756 person years; 906 direct and 856 induced). The HNAB losses, it is important to emphasize, are ongoing rather than temporary as in the case of the bridge construction jobs. Ideally, it would be possible to compare the HNAB losses with

[11] Recall that Smith et al.'s study period extended to 2035 while that here ended in 2024 due to the limited ability of MRIO's to project impacts.

the annual, ongoing job gains predicted for the bridge once it is in operation; however, no estimate of the latter could be found.

Another construction megaproject—the $7.3 billion twinning of the 1147 km Trans Mountain pipeline—provides an additional basis of comparison. According to documentation submitted to the National Energy Board (2016) by the project's proponent, the project would support 65,184 person years of direct, indirect, and induced employment over the first 20 years of operations, for an average of 3260 person years annually. This level of ongoing employment creation is roughly equal to the total annual employment losses found here for HNABs during 2020–2024 under business-as-usual (3024). It is about four times the size of the annual employment losses found to be avoided in the 2020–2024 period if policy action is taken to reduce P loadings (812 person years).

These comparisons confirm that the economic impacts associated with Lake Erie algal blooms are, indeed, on a scale sufficient to warrant provincial and national attention. At the same time, they remain relatively small in comparison to the overall provincial economy and, other things equal, within its capacity to absorb. This does not mean, however, that the regional impacts of HNABs will go unnoticed—especially in the more significantly impacted counties/regions like Haldimand County—and that many individuals do not stand, in one way or another, to have their economic well-being significantly impacted by the blooms.

Conclusion

The present study demonstrates that rigorous evaluation of the economic impacts of agriculturally related freshwater degradation is possible and that the size of the impacts—at least in the context of Lake Erie—is sufficient to warrant the attention of government policy makers. At the same time, the study revealed a variety of areas where both the physical and economic understanding of HNABs could be improved. While we believe our results are robust, we also believe their accuracy could be improved if such improvements were pursued.

To start with, the study could have benefited from a better biophysical basis for predicting the evolution of algal blooms. Smith et al. were obliged to develop their own, relatively simplistic, model of the evolution of HNABs on Lake Erie. It would have been preferable if a model of future blooms developed by scientists' experts in the lake and its ecology had been available for use.

Secondly, the economic understanding of algal blooms could be significantly improved if primary research were undertaken on the willingness-to-pay of residents of the Lake Erie basin for freshwater quality. As it was, Smith et al. relied on a meta-analytic regression model of willingness-to-pay (Johnston & Thomassin, 2010) that drew upon a variety of Canadian and US studies and was not focused specifically on the Laurentian Great Lakes. While the Johnston and Thomassin model is robust, without primary data on the willingness-to-pay of those living in the Lake Erie basin, we have no way of knowing the degree to which use of the model caused inaccuracies to our results. One reason to believe that the model may underestimate the economic losses associated with Lake Erie HNABs is that is in fact not measures of willingness-to-pay that would ideally be used in a study like ours, but measures of willingness-to-accept compensation for a loss in environmental quality.[12] Given that nonuser utility losses were found to drive the majority of the economic losses in this study, research in this area would seem a particular priority.

Thirdly, additional research on the impacts of HNABs on tourism and property values would be welcome. Smith et al. were obliged to make crude assumptions regarding the losses in tourism value added and property values due to HNABs. Both these categories, especially tourism, were the source of considerable economic losses in our estimates. Overall, we have the least confidence in these estimates.

Finally, research to better understand who the recreational users and nonusers of Lake Erie are would be welcome. Smith et al. assumed that all recreationists living within 50 km of the lake's shore would lose utility due to HNABs and that all individuals living within 100 km would lose nonuse utility. Given the size of Lake Erie and its status as an environmental "icon" these assumptions were thought to be conservative. Still, surveys to determine just how far individuals do regularly travel to enjoy recreation on Lake Erie and how far from the lake intrinsic concern about its quality can be assumed to extend would provide a firmer basis for the kind of work undertaken here.

Annex 1 Counties/Districts and Sectors Covered by REIM-Ontario

Sectors	Counties/districts	
1. Agriculture	1. Algoma	37. Peterborough
2. Fishing and hunting	2. Brant	38. Prescott and Russell
3. Logging and forestry	3. Bruce	39. Prince Edward
4. Mining	4. Chatham-Kent	40. Rainy River
5. Food and beverages	5. Cochrane	41. Renfrew
6. Rubber and plastic	6. Dufferin	42. Simcoe
7. Textile and clothing	7. Durham	43. Stormont, Dundas and glengarry
8. Sawmill industries	8. Elgin	44. Sudbury
9. Veneer and plywood	9. Essex	45. Thunder Bay
10. Other Wood Industries	10. Frontenac	46. Timiskaming
11. Furniture	11. Greater Sudbury	47. Toronto

[12] This is because our objective was to measure the losses people suffer when they lose access to ecological goods and services. For this, it would be more appropriate to use measures of willingness-to-accept than willingness-to-pay. The distinction is likely to be empirically significant, as willingness-to-accept is typically found to several multiples of willingness-to-pay (OECD, 2006)

Sectors	Counties/districts	
12. Pulp and paper Mills	12. Grey	48. Waterloo
13. Converted paper products	13. Haldimand	49. Wellington
14. Printing and publishing	14. Haliburton	50. York
15. Primary metals	15. Halton	
16. Metal fabricating	16. Hamilton	
17. Machinery and equipment	17. Hastings	
18. Motor vehicles	18. Huron	
19. Motor vehicle parts	19. Kawartha Lakes	
20. Other transport equipment	20. Kenora	
21. Computer and electronic equipment	21. Lambton	
22. Electrical equipment	22. Lanark	
23. Non-metallic minerals	23. Leeds and Grenville	
24. Petroleum products	24. Lennox and Addington	
25. Chemicals and chemical products	25. Manitoulin	
26. Other manufacturing	26. Middlesex	
27. Construction	27. Muskoka	
28. Transportation and storage	28. Niagara	
29. Communication and utilities	29. Nipissing	
30. Trade	30. Norfolk	
31. Finance, insurance, and real estate	31. Northumberland	
32. Business and professional services	32. Ottawa	
33. Educational and health services	33. Oxford	
34. Accommodation services	34. Parry Sound	
35. Other services	35. Peel	
36. Arts and recreation	36. Perth	

References

Coskun, Y., Atasoy, B. S., Morri, G., & Alp, E. (2018). Wealth effects on household final consumption: Stock and housing market channels. *International Journal of Financial Studies, 6*, 57.

Environment and Climate Change Canada and Ontario Ministry of the Environment and Climate Change (2018). *Canada-Ontario Lake Erie Action Plan—Partnering on achieving phosphorus loading reductions to Lake Erie from Canadian Sources*, Catalogue En164–54/2018E-PDF. Retrieved from https://www.canada.ca/en/environment-climate-change/services/great-lakes-protection/action-plan-reduce-phosphorus-lake-erie.html

International Joint Commission (2013). *Lake Erie ecosystem priority: Scientific findings and recommendations to reduce nutrient loadings and harmful algal blooms*. Retrieved from https://legacyfiles.ijc.org/tinymce/uploaded/Draft%20LEEP-Aug29Final.pdf

Isard, W. (1960). *Methods of regional analysis: An introduction to regional science*. The MIT Press.

Johnston, R. J., & Thomassin, P. J. (2010). Willingness to pay for water quality improvements in the United States and Canada: Considering possibilities for international meta-analysis and benefit transfer. *Agricultural and Resource Economic Review, 39*(1), 114–131.
Michigan Sea Grant. (n.d.). *Harmful algal blooms & Muck—What's the difference?*, on-line information sheet. Retrieved from https://www.michiganseagrant.org/wp-content/uploads/2018/10/HAB_Muck-2009.pdf
National Energy Board. (2016). *National energy board report: Trans Mountain expansion project*. Cat. No.: NE4-4/2016-3E. National Energy Board.
OECD. (2006). *Cost-benefit analysis and the environment—Recent developments*. Organisation for Economic Co-operation and Development.
Pichette, L. (2004). Are wealth effects important for Canada? Bank of Canada Review, Spring.
Smith, R. B. B., Bass, D. Sawyer, D. Depew, S. B. Watson (2019). Estimating the economic costs of algal blooms in the Canadian Lake Erie Basin. *Harmful Algae*, 87. https://www.sciencedirect.com/science/article/abs/pii/S1568988319300915?via%3Dihub
Winch, D. (1973). *Analytical welfare economics*. Penguin Press.
Winkler, S. (2016). Empirical evidence of wealth effects on consumption. Paper prepared for the 34th IARIW General Conference Dresden, Germany, August 21–27.
Workforce WindsorEssex (2019). Bridging our workforce: A guide to jobs related to the Gordie Howe International Bridge. Retrieved March 27, 2019, from, https://www.workforcewindsoressex.com/wp-content/uploads/2019/03/Bridge-Jobs-Guide-EN.pdf.

Chapter 4
Peri-Urban Agriculture and Food Security in MENA Countries

Alban Thomas

Abstract We present an analysis of the potential contribution of urban (UA) and peri-urban (PUA) agricultural systems to food and nutritional security in Middle Eastern and North African countries. We provide empirical evidence regarding the structure of UA and PUA that may be in favour of food and nutritional security in the MENA region, using several sources of information. The latter include a recent foresight study on future food systems in the MENA region in 2050, a brief literature review on UA and PUA systems, and an empirical analysis based on global data for MENA countries. We find evidence that the proportion of UA and PUA contributes positively to the city's capacity to feed its population, but there is no evidence of a size effect of peri-urban area (or its density), indicating that it may be more an intensive margin impact at play. Regarding the share of PUA in total peri-urban area, we find a positive and significant correlation with available food per head, while it is positive for food diversity, as well as a higher elasticity of animal-sourced protein contents of diet to GDP per capita in MENA countries on average. More data complemented with field surveys would be necessary to confirm our findings, but our empirical analysis covering all MENA countries confirms some assumptions found in the literature, regarding the positive contribution of UA and PUA systems for food and nutritional security at the country level.

Keywords Agriculture · Food security · MENA · Peri-Urban

A. Thomas (✉)
The Grenoble Applied Economics Laboratory (GAEL), UMR CNRS 5313/UMR INRAE 1215, Université Grenoble Alpes, Saint-Martin-d'Hères, France
e-mail: Alban.Thomas@inrae.fr

A. A. Kubursi et al. (eds.), *Urban Food Security in a Crisis Prone World*,
https://doi.org/10.1007/978-3-031-89440-4_4

Introduction

It is widely recognized that urban growth is a major contributor to the loss of agricultural soils in the Mediterranean and semi-arid countries, a phenomenon amplified by poor (of the absence of) urban planning policies. However, it is also true that urban areas, especially in developing countries, have a role to play in food and nutritional security, through, e.g., urban and peri-urban agriculture. As discussed by Zubel and Apprioual (2020), even though the natural way for food production lies in the countryside, there is a potential "for quickly growing cities and their suburbs to partially take over duties from rural partners, i.e., by producing herbs and vegetables (small-scale farms) in urban conditions." Unfortunately, the prospects of developing Urban Agriculture (UA) and Peri-Urban Agriculture (PUA) in world's regions such as MENA (Middle East and North Africa) require several necessary initial steps. First, these systems often suffer from a lack of recognition on behalf of policy makers in many developing countries, in particular when UA and PUA initiatives by dedicated international programmes have not been implemented at a sufficiently large scale. Second, the scientific literature on UA and PUA systems tends to be focused on industrialized countries and some selected world's regions, in particular sub-Saharan Africa and Southeast Asia. This relatively low level of scientific evidence for world's regions such as MENA has made it very difficult to evaluate the potential benefits from UA and PUA. This has also implied that cost-benefit analyses (useful for comparing policies to promote UA and PUA) cannot be conducted given the current state of scientific knowledge in the MENA region.

The objective of this chapter is to provide additional evidence regarding the potential of UA and PUA systems for food and nutritional security in MENA countries, using three different sources of information. First, foresight studies have addressed the question of food security by taking into account transitions in food systems and land use changes under climate change scenarios. Although not explicitly focused on UA and PUA, these studies are useful in identifying possible drivers that may be in favour of the expansion of UA and PUA in MENA countries. This is particularly true of a recent foresight study conducted by INRA, France (Le Mouël & Schmitt, 2018). Second, even if foresight studies focusing on UA and PUA systems in the MENA region are virtually not represented in the scientific literature, there is now a significant number of review and meta-analyses that address the benefits and factors behind adoption of UA and PUA systems in developing countries. As some have similarities with MENA countries, it is useful to explore, with a qualitative and comparative approach, the main results from these studies. Third, some datasets at the global level are made available on MENA countries and include some aspects of UA and PUA, and they can be used to conduct preliminary empirical analyses. Data at the global scale are obviously far from being enough to provide a comprehensive description of UA and PUA systems, for which field surveys and qualitative or mixed-methods approaches are necessary. However, for a first empirical step covering all MENA countries, the use of global-level data may be used to start and explore the validity of assumptions found in the literature regarding the contribution of UA and PUA to food and nutritional security.

The outline of the chapter is as follows. In the first section, we present the main challenges facing agricultural and food systems in MENA countries, based on a recent foresight study conducted by INRA in 2018. This section contains a summary of the scenarios that are discussed in this foresight study, regarding food security in the MENA region, with a particular attention to scenarios that may be beneficial for the development of UA and PUA. Second, based on recent literature reviews on benefits and challenges associated with UA and PUA, we explore the potential of such agricultural systems for MENA countries, given the characteristics of the latter. In the third section, we test several assumptions that are frequently found in the literature on UA and PUA, on a sample of MENA cities. Restricted to PUA, these assumptions concern the relationship between households' food security and the structure of PUA, as well as the relationship between PUA and food and nutritional security at the country level. For lack of adequate data, other contributions of UA and PUA, such as their role in women empowerment and poverty reduction could not be discussed.

Food Security in the MENA Region: A Foresight Study

An exhaustive foresight study has been conducted by INRAE (French Research Institute for Agriculture, Food and the Environment, then INRA in 2018) researchers in France, on the state of food security in the MENA region, exploring present challenges facing MENA countries in the face of transitions in food systems and global change (Le Mouël & Schmitt, 2018). Starting with a retrospective statistical analysis to evaluate trends in food consumption, agricultural productivity and land use, the authors then explore the state of food security in the future (2050), using a mixed method combining partial-equilibrium modelling and qualitative scenario design and assessment. The following is a summary of their assumptions, scenario building and main results. Although there are similarities in terms of climate, use of some natural resources and urbanization patterns, a first step is to acknowledge the diversity of MENA countries in terms of agricultural potential, land use and even diets. For this reason, it is important to distinguish several 'blocks' composing the MENA region: Egypt, North Africa,[1] Middle East,[2] Near East[3] and Turkey.

Looking at the MENA region as a whole, a major challenge is the high dependency on agricultural and agri-food imports (around 40% of total demand for food at the end of last decade), especially cereals, while demand for food is increasing rapidly due to demographic growth, urbanization and changes in diets. Because of

[1] Algeria, Morocco, Tunisia, Libya and Mauritania.

[2] Iran, Iraq, Saudi Arabia, Kuwait, United Arab Emirates, Yemen, Bahrain, Qatar and Oman.

[3] Syria, Lebanon, Jordan, Israel, Palestinian Territories.

agronomic and climate constraints, and of the limited impact of agricultural policies, agricultural production is not increasing at the same pace, possibly leading to an even increasing dependency on imports. According to the authors, such dependency could reach 50% by the year 2050 on average (from 31 to 53 percent for Egypt, from 51% to 63% for Near East and Middle East, while remaining roughly unchanged for North Africa and Turkey). However, because of climate change, dependency on imports would increase even more, to reach 53% in 2050 and affect North Africa in particular (from 54% to 69% in 2050 under significant climate change). Because of its hydrological and geographic situation, Turkey could see its arable land increase and confirm a position of net exporter of agricultural and food products by 2050. Moreover, a reduction in the agricultural potential of the region would not necessarily translate into increasing imports of plant products, but rather an increase in animal products imported to the region to compensate for the decrease in local livestock production. Another consequence of climate change in MENA countries would also be a drop in agricultural productivity of labour (about 11% on average from its projected trend, except for Turkey), leading to a growing discrepancy between rural and urban average incomes and a maintained poverty level in rural areas (Le Mouël & Schmitt, 2018).

Regarding urban areas, the MENA region is characterized by a strong growth of its urban population (multiplied by 6 over the past 50 years, with more than 65% of population now living in urban areas). To cope with the low average level of precipitation (except in Turkey and Iran) and a significant intra-year variability of rainfall, policies for developing irrigation have been implemented, leading to a proportion of about 34% of crops being irrigated (almost 100 percent for Egypt, 55% for the Middle East, about 11% in North Africa). However, irrigation efficiency is generally low because of poor management of water networks and irrigation relying mainly on gravity-based methods (furrow, etc.). More than 80% of all renewable water resources are already in use in the region (and even beyond in the Arabian Peninsula), leading to increasing competition with other water uses (residential, industrial, energy). This is a major concern for urban and peri-urban development, as well as for agricultural systems in such areas, if urban water networks are not capable of providing an adequate volume level and service quality (pressure) for residential and urban and peri-urban agricultural users.

It is particularly interesting to discuss the projected trends in households' diets for the year 2050, obtained from a retrospective (hindsight) statistical analysis. Under the authors' assumptions, the average diet of the MENA region would be characterized by a much lower proportion of wheat in plant food products (about 35%t), compensated by an increase in maize, oilseed and protein crops, and fruits and vegetables. Regarding animal products, the expected future diet would experience an ever-increasing importance of poultry meat, reaching one third of demand for animal products, and of fish (about 8.5% of this demand), while the share of milk and dairy products would remain limited (about 40% of animal products in 2050).

Turning now to the exploration of scenarios for regional food systems, the foresight study considers four alternative scenarios that are likely to modify the dependency on agricultural and food imports in the region: (a) "westernization" of diets ("Western diet"); (b) technical change; (c) Mediterranean diet; (d) reduction in losses and waste. The first scenario would lead to an increase in the demand for animal products, oil and fats and sugar at the expense of cereals and fruits and vegetables, with a stronger dependency on imports as a result. Naturally, it is expected that such scenario would be mostly driven by the growth of population in urban and peri-urban areas. An important consequence of the "Western diet" scenario is the strong increase in demand for pasture, mostly in Middle Eastern countries, which would not be met with local livestock systems, unless a major improvement in breeding and productivity, in general, is achieved. In the Mediterranean diet scenario, the regional demand for food would decrease, especially the demand for animal products, oils and fats and sugar (by contrast with the "Western diet" scenario), with a limited impact on import dependency. This is also true for scenarios b) and d), associated with more productive and efficient agricultural and food processing systems, achieving higher productivity levels and economic value per unit of land, water use and other production inputs. Furthermore, none of the sets of possible actions, taken individually, would allow compensating for the expected dependency on agricultural and food imports of the MENA region. Combining the technical change, Mediterranean diet and the reduction in losses and waste scenarios would allow for a better adaptation to climate change and a stabilized dependency on imports. They would, however, require ambitious and well-designed public policies to foster and promote research and development, investment in irrigation infrastructure, technical training, nutritional policies, etc.

It is clear that the projected trends in the composition of average diets in the MENA region, possibly accompanied by scenarios for reducing dependency on agricultural and food imports, provide interesting prospects for UA and PUA. More precisely, the increased demand for fruits and vegetables and poultry meat would benefit production from UA and PUA systems, which match their very low requirements in terms of land, pasture and inputs used in field crops (fertilizer, machinery). The increase in demand for fish is also a possible outlet for urban and peri-urban systems with coastal areas. The Mediterranean scenario relying more on local food, in particular on fruits and vegetables and less animal products, would also be in favour of UA and PUA, while reducing food waste and losses is clearly associated with urban food systems based on, e.g., the circular economy and a better management of biomass produced and transformed in urban areas.

Reviews on Urban and Peri-Urban Agricultural Systems

As mentioned above, the scientific literature on UA and PUA systems has rarely produced analyses focused on MENA countries, while the bulk of research tends to be on industrialized countries and some of the world's regions, in particular sub-Saharan Africa and South-East Asia. As noted by Orsini et al. (2020), research on UA and PUA in developing countries has mainly targeted their contribution to food security, while in developed countries, it is the combination of services provided by agriculture (including; however, social and poverty-reduction objectives for economically vulnerable households) and the associated business models that have attracted the attention of researchers. Nevertheless, it is interesting to gather a selection of findings that may be relevant for MENA countries, in particular when agricultural structures, climate and geographic, or social and economic conditions are similar. The following is a summary of the findings of literature reviews addressing UA and PUA systems in general, irrespective of their economic development, or in developed countries (the "Global North").

Orsini et al. (2020) explore the multifunctional nature of UA in the 'Global North', through its contribution to ecosystem services provisioning (food provision, health functions, social inclusion and justice, contribution to ecological and environmental sustainability, etc.), and identify the factors that either hinder or promote the regional diffusion and uptake of UA. Beyond a description of the typology of UA systems and their associated business models, they discuss opportunities for crop diversification in the urban environment. The classification proposed by the authors is applied to a sample of 470 UA projects in the Global North. The main factors behind the development of UA include the legal framework, access to land, contamination risks, local climate conditions and availability of natural resource. The typology of UA farming systems includes allotment of urban gardens, extensive peri-urban farms, urban community gardens, rooftop farms and indoor vertical farms.

Dorr et al. (2021) propose a literature review and a meta-analysis of the impacts of UA on natural resource use and the environment. Starting from the finding that the literature on UA systems has produced ambiguous results regarding their environmental performance, the authors perform a systematic review of life cycle analyses (LCA) of UA systems (88 different farms and 259 production systems). The literature review reveals that the majority of case cases are in Europe (60%), followed by Asia (20%), North America (8%), South America (6%), Australia (4%) and Africa (2%). No analysis using LCA on UA system in MENA countries was found in the literature. Most studies were done in Spain (31%), China (8%), Italy (8%), France (8%) and the United States (8%), and the cities most studied were Barcelona (25%) and Beijing (8%). Interestingly, these cities are characterized by hot summers, and in the case of Barcelona, warm temperate on average. Environmental performance of UA systems is compared in terms of water consumption, greenhouse gas emissions and cumulative energy demand. Results from the meta-analysis reveal that there is a very wide range of system types, crop choices and economic orientations, and that UA has a potential for both extremely small and large impacts and yields. The authors found that UA can substitute conventional agricultural without increasing food system impacts. Unfortunately, the quality and consistency of the LCAs were not sufficient, and the variability in results reviewed was too high, preventing the authors from drawing definitive conclusions about the environmental impacts and resource use of UA systems.

Clinton et al. (2018) conduct a global analysis by country, of ecosystem services provided by UA, namely, food production, energy savings, nitrogen sequestration, avoided storm water runoff, pollination, climate regulation soil formation and biological pest control. Using a wide range of data (including data from FAO, meteorological, terrain, urban, climate, etc.), they estimate that UA may contribute to a grand total of between 80 and 160 billion USD annually, if implemented in an intensive way. The authors confirm the large heterogeneity across countries, in terms of contribution to food security and to the provision of ecosystem services by UA systems. Yan et al. (2022) analyse a collection of 605 scientific papers from the Web of Science between 2001 and 2021, and using a visualization mapping tool, evaluate the developments on urban agriculture. They find that the number of publications on UA increased substantially every year during that period, with the Unites States, Germany, the United Kingdom, Italy, and China as the most influential countries on the subject. Research on urban agriculture focuses not just on food production but also on the various functions and services provided by UA, related, for example to the water-energy-food nexus. It also turns out that UA is not necessarily a synonym for resource-conserving and environmentally friendly food systems.

Martellozzo et al. (2014) acknowledge the general view that UA may help reduce food insecurity and environmental impacts of agricultural activities, by producing food in high population density areas, thereby reducing transportation costs and connecting consumers directly to food systems through an efficient use of urban area structure. The authors evaluate the global amount of food that could be produced by UA systems, using several global-scale datasets. They find that about one-third of the total global urban area would be required to meet the demand for vegetables of the urban population worldwide. However, their data does not allow them to determine what would be the proportion of urban areas suitable for UA, which is expected to vary substantially across countries (and across UA systems). Significant deviations from the global average production value are mostly due to variations in urban population density, and much less to variations in agricultural yields or food consumption per head. Unfortunately, in more food-insecure countries, the pressure on urban area is highest where UA is also most needed. The authors also show that about two-thirds of UA expansion may be due to small urban clusters, implying that UA promotion policies should also target smaller urban areas.

In what follows we present the main results of a series of reviews that focus on developing countries (also labelled 'Global South'). Follmann et al. (2021) perform a systematic review of UA and PUA systems under rapid urban growth in Global South cities, using a meta-analysis of 92 scientific papers on 83 cities. They focus on the interactions between urban growth, agricultural production and urban food markets, to show that the most frequent challenges facing UA and PUA concern agricultural land conversion and spatial expansion of urban areas. However, some studies indicate that increasing food demand with a modification of the structure and composition of diets can also explain the spatial expansion and intensification of UA and PUA. Their analysis confirms that farmers adapt to urban growth in multiple ways, posing a multidimensional challenge to urban planners and policy makers involved in food and agriculture. Few studies in their meta-analysis address more complex patterns such as spatiotemporal dynamics, intra-urban variations and multidimensional interlinkages of UA and PUA in a context of urban growth. The authors conclude by the necessity to use mixed-methods, linking global data (remote-sensing, geospatial and global) to qualitative and quantitative fieldwork data, to better understand the dynamics of social and spatial changes of UA and PUA in Global South cities.

Orsini et al. (2013) review the many factors (social, cultural, technical, economic, environmental, political) affecting UA in selected case studies from East Asia, South America and East Africa. They address the benefits and limitations of UA, to show that, in particular, food security benefits of UA are provided to between 100 and 200 million urban farmers worldwide, mostly from fresh horticultural goods. Social improvement is also a benefit from UA, in countries where poor households (including most urban farmers) may spend more than 75 percent of their income on food. UA has also a positive effect by reducing gender inequalities, 65 percent of urban farmers being women. In terms of environmental impacts, the authors show that UA has benefits for environmental conservation by reducing urban solid waste, improving urban biodiversity and air quality, and reducing environmental impacts associated with food transport and storage.

De Bon et al. (2010) discuss the main functions of UA, i.e., its social roles and economic functions as part of its multifunctionality, but also the constraints and risks for human consumption and the environment. They claim that UA will continue to contribute significantly to urban households' income, with differences between rural and urban households decreasing over time. UA systems include aquaculture, livestock and plants, the most common crops being perishable leafy vegetables, particularly in Southeast Asia and Africa. In terms of differentiation, vegetable products have shorter marketing chains with lower price differentials between farmers and consumers than longer supply chains. UA production faces, however, various challenges: access to agricultural inputs (including fertilizer and water) and the necessary limitation of its negative environmental impact. Even though UA can reuse urban waste, this additional source of biomass will not be enough to achieve high agricultural yields, while introducing the risk of producing contaminated food products.

Hamilton et al. (2014) perform a literature review of UA in developing countries, and they identify several dimensions of urban agriculture that require a better understanding. These dimensions include the global and regional extent of UA, contribution of UA to communicable diseases, the role of UA in abating malnutrition and obesity, the impacts of UA on women and the risks associated with chemical inputs.

Zezza and Tasciotti (2010) acknowledge the fact that detailed data on a large number of UA systems in developing countries are rare, while most survey data on UA are qualitative, which makes quantitative comparisons problematic. The authors use nationally representative household survey data for 15 developing or transition countries, to assess the role of UA in reducing poverty in urban areas. They show that the role of UA in reducing urban poverty and food insecurity is there, but it should not be overemphasized. This is because the share of UA in income and in total agricultural production is often limited. However, UA may be particularly promising in much of Africa and in countries where agriculture provides a substantial share of income for the urban poor. The authors also find evidence of a positive statistical association between involvement in UA agriculture, and that engagement in farming in urban areas is positively associated with higher calorie availability and greater diet diversity. Unfortunately, no MENA countries are included in their analysis (the 'closest' countries to the region being Bangladesh and Pakistan). Overall, their findings indicate that agriculture is a significant activity in the urban economy of most developing countries, involving between 10 and 70 percent of urban households. However, the role of UA on income generation seems to be much more limited, with the important exception of the African countries, where poor households are largely over-represented in UA in all regions.

Finally, we summarize two research papers that propose case studies in MENA countries, without a comprehensive and global analysis of the relationship between UA and PUA, on the one hand and food security on the other. Nevertheless, they are useful in understanding the special features UA systems may have in the region, possibly with factors facilitating the development of UA and PUA.

Tohmé-Tawk et al. (2014) discuss the constraints faced by sustainable UA systems in developing countries, and they present a multi-stakeholder approach implemented in Amman (Jordan) in the context of a project led by RUAF Foundation (Resource Centres Network on Urban Agriculture and Food Security) to mainstream urban agriculture in the MENA region. A Participatory and Multi-stakeholder Policy formulation and Action Planning (MPAP) process was used to design a policy narrative and identify priority issues that included improving access to water by reuse and rainwater harvesting; supporting UA by providing adequate training and extension services; introducing adequate laws and regulations in favour of UA; supporting farmers' markets and fairs and improving post harvesting techniques.

Lteif and Soulard (2019) present the case of Beirut to illustrate adaptation of UA to land tenure, focusing on *waqf* land (or *habous* land), whose revenue benefits the community in the form of pious actions. The authors identify three main forms of UA in Beirut: relic agriculture, farming where *waqf* land allows multiple uses, and urban breeding. They show that *waqf* land provides greater security for farmers, while other private lands used for agriculture are characterized by higher land tenure insecurity. In their paper, the authors also discuss the general trend of MENA countries South and East of the Mediterranean Sea, where food security considerations led policy makers to place their priorities on cereal imports, especially during and after the 'food crisis' of 2007–2008. Even though this crisis revealed the fragility of the global food system (and the potentialities of UA in these countries for reducing instability in access to food), local governments were more concerned about urban planning and the priority for urban dwellers. However, UA seems to maintain itself in Arab Mediterranean cities, but often with an 'invisible status' (example of dairy farmers in Cairo and small-scale urban horticulture in Meknes).

Empirical Application: Peri-Urban Agriculture in Selected MENA Cities

The objective of this section is to test several assertions that are frequently found in the literature on UA and PUA, on a sample of cities in the MENA region. Restricted to PUA (not UA, for lack of data) these assertions concern the (positive) relationship between food and nutritional security, on the one hand and the structure of PUA, on the other. For lack of adequate data, the role of PUA development in women empowerment and poverty reduction could not be discussed.

We collected data from the Potsdam Institute for Climate (PIK database CI-GRASP[4]) on PU and PUA geographical features and carrying capacities (agricultural potential of UA and PUA areas) for 70 cities in the MENA region in the year 2020. Data was only available from PIK for large-scale cities, ranging from 333,000 (Suhaj, Egypt) to 11 million inhabitants (Cairo, Egypt). Our dataset was complemented by country-level data from FAOSTAT (FAO) on average diet and food and nutritional security, urban population growth, etc., for the latest year available (in most cases, 2020 or 2019). Table 4.1 presents descriptive statistics on our sample. On average, MENA countries have a significant rate of annual growth of their urban population, about 4.25 percent per year with a maximum of 8.5 percent. To feed this growing population, PUA areas in the largest cities have a carrying

Table 4.1 Descriptive statistics

Variable	Source and definition	Unit	Mean	Std. deviation	Min	Max
P_AGRI_PU	PIK. Share of peri-urban area devoted to agriculture	Proportion	0.2866	0.2870	0.0	0.9876
CARR_CAP	PIK. Average carrying capacity (percent of population potentially supplied by local PUA)	Proportion	0.2158	0.2254	0.0006	0.9972
PU_AREA	PIK. City's peri-urban total area	1000 ha	144.48	180.59	34.90	1101.70
P_PU	PIK. Peri-urban area over total artificial (urban and peri-urban) area	Proportion	0.2424	0.1993	0.0375	0.7706
GDP_HAB	FAOSTAT. GDP per head.	1000 USD/year	21.63	19.13	1.2490	85.26
GROWTH_URBAN	FAOSTAT. Annual growth rate of country's urban population	Growth rate (last 2 available years)	1.0425	0.0112	1.0169	1.0852
AVAIL_FOOD_HAB	FAOSTAT. Available food per head in country	kcal/person	3212.77	375.48	2012.0	3742.0
P_CEREAL	Share of cereals, roots and tubers in country's average diet	Proportion	0.5517	13.2042	34.0	92.3
P_PROTEIN	Quantity of proteins available in average diet	FAOSTAT. g/person/day	88.5971	17.75	52.0	125.0
P_ANIMAL	Quantity of available proteins of animal origin in average diet	FAOSTAT. g/person/day	34.6514	19.78	9.0	92.3

$N = 70$. Source: PIK CI-Grasp Data and FAOSTAT

[4]CI-GRASP: Climate Impacts—Global and Regional Adaptation Support Platform, version 2.0. Potsdam Institut für Klimafolgenforschung (PIK).

capacity of about 21% on average (variable CARR_CAP), representing on average 28% of total peri-urban area, which itself counts for about 24% of rural artificial area (variable P_PU).

As discussed above, MENA countries have divergent profiles in terms of land use, agricultural productivity and food security. For this reason, it is useful to produce statistics on food availability, PUA and PU extent for five MENA sub-regions: Maghreb (North Africa), Egypt, Near East, Middle East and Turkey. These statistics for the carrying capacity of PUA, the ratio PUA over total PU area and the ratio PU over total artificial area (urban + peri-urban) are presented in Table 4.2.

The first step in our empirical analysis is to explore the statistical relationship between geographical characteristics of cities in our sample, and their carrying capacity (in terms of food provision to the local population). Using the PIK data at the peri-urban level, we estimate a regression equation of the carrying capacity of PUA as a function of the fraction of peri-urban area dedicated to agriculture. This expected effect of "agricultural density" can be interpreted as an indicator of land sharing in peri-urban areas in MENA countries' largest cities. We also include two additional covariates, which are tested separately from P_AGRI_PU to check for robustness in the full equation: PU_AREA (size of peri-urban area) and P_PU (ratio of PU area over urban and peri-urban areas), in Models (I) and (II). Parameter estimates are presented in Table 4.3. When variable P_AGRI_PU is not included, the two control variables PU_AREA and P_PU are not significant, so that there does not seem to be a size effect (PU_AREA) of a PU density effect (P_PU) on carrying capacity. On the other hand, P_AGRI_PU is positive and significant in the full

Table 4.2 Mean values for selected indicators, by MENA sub-region

Sub-region or country	Carrying capacity of PUA (%)	PUA/total PU area (%)	PU/total artificial (urban + peri-urban) area (%)
North Africa	26.34	41.11	25.41
Egypt	54.90	60.41	19.22
Near East	21.12	34.39	48.65
Middle East	6.17	7.78	23.25
Turkey	31.29	33.58	9.50

N = 70. Source: PIK CI-Grasp Data

Table 4.3 Parameter estimates. Carrying capacity of PUA

Variable	(I)	(II)	(III)
Intercept	22.4757*** (3.1692)	15.1835** (5.7408)	1.1853 (3.1082)
P_AGRI_PU	–	–	69.6374*** (5.7582)
PU_AREA	−0.0061 (0.0121)	−0.0428 (0.0283)	−0.0024 (0.0147)
P_PU	–	51.9807 (36.8453)	0.7015 (19.1214)
R2	0.0280	0.1669	0.8509

N = 70. Estimation method: fixed effects (country). Robust standard errors are in parentheses. *, ** and *** respectively denote parameter significance at 1, 5 and 10 percent. (i): in 1000 ha

equation—model (III)—confirming the assumption that the proportion of agriculture in peri-urban areas contributes positively to the city's capacity to feed its population. Moreover, the absence of any size effect of peri-urban area (or its density) tends to indicate that it is more an intensive margin impact (more agriculture on the same area) that is at play, and not an extensive margin effect (more hectares with the same agricultural output per hectare), on food supply per head.

The second step of the empirical analysis consists in testing whether there is a significant contribution of PUA to food security and diet diversity at the country level, as it is mentioned as a positive aspect of such systems (see above). We are aware of the limitations of this analysis, which combines data at the city level and country data, so that empirical results should be considered with care. In particular, consistency of estimates relies on the assumption that urban and peri-urban areas in our sample are representative of such areas at the country level. Given the very limited number of observations and that our sample only consists of the country's largest cities, one should interpret our empirical question as follows: is there a significant contribution of the country's PUA from the biggest cities to food and nutritional security, the latter being measured by available food per head and diet composition at the national level?

Estimation results are presented in Table 4.4 where, to limit the number of covariates to a minimum, we consider only GDP per capita as a control variable in four equations of food and nutritional security at the country level: available food (kcal/day) per head, share of cereals, tubers and roots in diet, protein contents of diet and animal-sourced protein contents of diet. We estimate all four equations in log-log form to obtain elasticities directly from parameter estimates. The share of PUA in total peri-urban area (P_AGRI_PU) is positive in all equations, but only significant at the 5% level in the equation for available food per head. Average wealth per inhabitant (GDP_HAB) is positive and significant in all equations except for the share of cereals, roots and tubers. We find a higher elasticity of animal-sourced protein contents of diet to GDP (0.41), compared with protein contents (0.18) and available food per capita (0.13), reflecting a larger income elasticity for animal

Table 4.4 Parameter estimates—Food availability and diet indicators

Variable	Available food per head	Share of cereals, roots and tubers in diet	Protein contents of diet	Animal-sourced protein contents of diet
Intercept	0.8892*** (0.0718)	4.1677*** (0.1309)	4.0805*** (0.0894)	2.5962*** (0.2218)
P_AGRI_PU	0.0622*** (0.0226)	0.0270 (0.0394)	0.0593* (0.0308)	0.1249* (0.0662)
GDP_HAB	0.1350*** (0.0229)	−0.0714 (0.0593)	0.1791*** (0.0316)	0.4134*** (0.0857)
R2	0.6412	0.2521	0.5100	0.5542
RMSE	0.1130	0.2205	0.1916	0.1916

$N = 17$. Robust standard errors are in parentheses. *, ** and ***, respectively, denote parameter significance at 1, 5 and 10 percent

products on average. Although, in terms of diet diversity and composition (cereals, roots and tubers, proteins, animal proteins), the statistical relationship with the proportion of PUA is not significant at the 5% level, we can still interpret the relative magnitude of parameters. Moreover, the contribution of PUA in relative terms (P_AGRI_PU) to diet diversity is higher in the case of animal-sourced proteins (0.12) than for protein contents (0.06) and food availability (in kcal/head) in general (0.06). The fact that the share of cereals, tubers and roots is not significant is also consistent with the fact that, as discussed above, PUA is less suited for field crops. However, this interpretation is only based in a very limited number of observations, and more data complemented with field surveys at the urban and peri-urban level would be necessary to confirm our findings.

Conclusion

This chapter has presented an analysis of the potential contribution of urban and peri-urban agricultural systems to food and nutritional security in Middle Eastern and North African countries. In the light of the limited data and scientific evidence regarding PU and PUA in these countries, this chapter provides additional evidence regarding the structure of UA and PUA that may be in favour of food and nutritional security in the MENA region. We used three different sources of information to achieve this goal: a recent foresight study on future food systems in the MENA region in 2050, a brief literature review on UA and PUA systems, in particular in developing countries, and an empirical analysis based on global data for MENA countries. Such a preliminary empirical analysis is based on a very small dataset, and estimation results should be considered with caution, because of data limitation (both in volume and level of detail).

In particular, we find evidence that the proportion of agriculture in peri-urban areas contributes positively to the city's capacity to feed its population. However, we do not see any size effect of peri-urban area (or its density), indicating that it may be more an intensive margin impact at play, and not an extensive margin effect (more hectares with the same agricultural output per hectare), on food supply per head. Regarding the share of PUA in total peri-urban area, we find a positive and significant correlation with available food per head, while it is positive but less significant for food diversity (as measured by the share of cereals, roots and tubers, proteins and animal-sourced proteins in the average diet). We also find a higher elasticity of animal-sourced protein contents of diet to GDP per capita in MENA countries on average, compared with protein contents and available food per head, reflecting larger income elasticity for animal products on average. More data complemented with field surveys at the urban and peri-urban level would be necessary to confirm our findings. Nevertheless, our empirical analysis covering all MENA countries allows us to confirm some assumptions found in the literature, regarding the positive contribution of UA and PUA systems for food and nutritional security at the country level.

References

Clinton, N., Stuhlmacher, M., Miles, A., Uludere, N., Wagner, M., Georgescu, M., Herwig, C., & Gong, P. (2018). Global geospatial ecosystem services estimate of urban agriculture. *Earth's Future, 6*, 40–60.

De Bon, H., Parrot, L., & Moustier, P. (2010). Sustainable urban agriculture in developing countries. A review. *Agronomy for Sustainable Development, 30*, 21–32.

Dorr, E., Goldstein, B., Horvath, A., Aubry, C., & Gabrielle, B. (2021). Environmental impacts and resource use of urban agriculture: A systematic review and meta-analysis. *Environmental Research Letters, 16*, 093002.

Follmann, A., Willkomm, M., & Dannenberg, P. (2021). As the city grows, what do farmers do? A systematic review of urban and peri-urban agriculture under rapid urban growth across the Global South. *Landscape and Urban Planning, 215*, 104186.

Hamilton, A. J., Burry, K., Mok, H.-F., Barker, S. F., Grove, J. R., & Williamson, V. G. (2014). Give peas a chance? Urban agriculture in developing countries. A review. *Agronomy for Sustainable Development, 34*, 45–73.

Le Mouël, C., & Schmitt, B. (Eds.). (2018). *Food dependency in the Middle East and North Africa region: Retrospective analysis and projections to 2050* (1st ed.). Springer Verlag. ISBN-13 : 978-9402415629.

Lteif, C., & Soulard, C. (2019). L'agriculture urbaine à Beyrouth, Liban. Quel rôle du foncier waqf ? *Rives méditerranéennes, 59*, 183–205.

Martellozzo, F., Landry, J.-S., Plouffe, D., Seufert, V., Rowhani, P., & Ramankutty, N. (2014). Urban agriculture: A global analysis of the space constraint to meet urban vegetable demand. *Environmental Research Letters, 9*, 064025.

Orsini, F., Kahane, R., Nono-Womdim, R., & Gianquinto, G. (2013). Urban agriculture in the developing world: A review. *Agronomy for Sustainable Development, 33*(4), 695–720.

Orsini, F., Pennisi, G., Michelon, N., Minelli, A., Bazzocchi, G., Sanyé-Mengual, E., & Gianquinto, G. (2020). Features and functions of multifunctional urban agriculture in the Global North: A review. *Frontiers in Sustainable Food Systems, 4*, 562513. https://doi.org/10.3389/fsufs.2020.562513

Tohmé-Tawk, S. T., Moussa, Z., Hamadeh, S. K. (2014) Mainstreaming Urban Agriculture in the Middle East and North Africa: a multi-stakeholder approach. 11th European IFSA, Berlin, Germany, 1-4 April, 2014, p. 1532-1538.

Yan, D., Liu, L., Liu, X., & Zhang, M. (2022). Global trends in urban agriculture research: A pathway toward urban resilience and sustainability. *Land, 11*(117). https://doi.org/10.3390/land11010117

Zezza, A., & Tasciotti, L. (2010). Urban agriculture, poverty, and food security: Empirical evidence from a sample of developing countries. *Food Policy, 35*(2010), 265–273.

Zubel, K., & Apprioual, A. (2020). Agriculture and food security in climate sensitive areas in the Mediterranean. Report for the European Committee of the Regions, Commission for Citizenship, Governance, Institutional and External Affairs. European Commission, Brussels. 56 p.

Chapter 5
Self-Provisioning and Urban Agriculture in Canada

Rod J. MacRae

Abstract Self-provisioning is an important, but weakly valued, part of the Canadian food system. This chapter provides a brief overview of the state of self-provisioning in urban areas of Canada and then proposes, using a transition framework, a range of strategies to increase self-provisioning and domestic food security.

Keywords Canadian food system · Self-provisioning · Urban agriculture

Introduction

For much of human history, most of us were hunters and gatherers and a bit more recently, farmers and gardeners. Survival was linked to the ability to feed self, family, and community. Sharing and later barter were important parts of the exchange process.

Specialization of tasks is a very recent phenomenon, as is the use of currency as a mediator of exchange and storehouse of value. As the food system has been transformed, it has allowed people to stop self-provisioning and use their earnings to purchase from those who are food producers, processors, distributors, retailers, and restauranteurs. The urbanization of the Canadian population has also been a significant force reducing self-provisioning. Food was pushed from the list of urban priorities with changes in food production practices in the late nineteenth century that made them appear, if not actually be, incompatible with urban density and key urban planning objectives. This transformation was part of the second industrial revolution, where food production and processing became inherently loud, pungent, messy, and even toxic (Patel, 2012), and mostly a rural activity, although many cities retain historical elements of an earlier time when food production and processing was common in cities. Some have viewed all this as liberation from the drudgery

R. J. MacRae (✉)
Faculty of Environmental and Urban Change (retired), York University, Toronto, ON, Canada
e-mail: rmacrae@yorku.ca

A. A. Kubursi et al. (eds.), *Urban Food Security in a Crisis Prone World*,
https://doi.org/10.1007/978-3-031-89440-4_5

of survival, others as a disaster leading to individual and community vulnerability. The heightened interest in gardening during COVID-19 speaks to this latter position.

Clearly most Canadians are now dependent on the market and the food system to acquire a nourishing diet. By definition, self-provisioning takes place outside the market and that is the primary reason it has declined in significance and is not supported by firms or the state. When self-provisioners are able to capture, harvest, preserve, slaughter, and butcher food themselves, traditional food system actors do not make money.

There are major impediments to increasing self-provisioning: loss of commons and commons resource management, loss of skills, contaminants, lack of access to land, inputs, equipment, community storage, slaughter and butchering facilities, food safety rules, and negative cultural perceptions. Benefits that can arise from an increase include: community cohesion and safety, recreation, greater cultural expression, reduced social isolation, better nourishment, improved spirituality, reduced costs, and ultimately better health and well-being. Foods acquired through self-provisioning are less expensive and frequently more nourishing than supermarket foods.

The State of Self-Provisioning in Canada

Also significant is the lack of data collection on self-provisioning, since most food system data collection focuses on economic dimensions. In Canada, we don't have very good data on the volume of vegetables consumed out of home gardens, or how much of the diet is provided from hunting and gathering. In particular, we don't have a clear picture of who relies on self-provisioning to assure an adequate diet, a critical situation if we're to do a better job of supporting such individuals, families, and communities.

Equally significant, it's not clear how important self-provisioning could be with the right supports in place to overcome the obstacles to better management. However, one indication of the potential comes from emergencies, such as the British WWII experience, when about 10% of domestic production (primarily fruit, vegetables, chickens, and hogs) came from home gardens and allotments inspired by the very extensive supports provided by the state, and some percentage of the population supplemented their diet with small game (Boyle, 2022).

Private Gardens

It appears that COVID caused more people to garden. A survey reported that half the residents harvested at least one fruit or vegetable from their garden in 2020, and of these 17% had started because of the pandemic. About 80% lived in single-family dwellings, but some 18% reported growing food in containers (Agri-Food Analytics

Lab, 2020). This suggests that pre-pandemic, about one-third of Canadians had been food gardening. Statistics Canada (2013) had reported that 57% of Canadians gardened in 2013 but their analysis included flowers, so it is not clear how many were just gardening for food. Almost twice as many home owners reported gardening than renters. A Montreal analysis reported that 40% of its household had gardens, amounting to 68–208 ha devoted to food and nonedible plants, producing an estimated 2200–6600 tonnes of vegetables with a retail market value of $15–48 million (Duchemin & McClintock, 2020). How important gardening is for self-provisioning, nutritional health, and finances is unclear, though many studies from North America report that gardening saves hundreds of dollars per family in supermarket expenses and provides nourishing food and higher consumption of fruits and vegetables (cf. Alaimo et al., 2008; Litt et al., 2011; CoDyre et al., 2015; Duchemin & McClintock, 2020).

Community Gardens and Greenhouses

In early periods of the twentieth century, many Canadian cities had extensive areas devoted to gardens, including greenhouses, but much of that infrastructure was dismantled and is only slowly being rebuilt. In Toronto in 1915, some 2000 garden plots coordinated by the Rotary Club generated almost $one million in produce in current dollars (Johnson, 2009). In 1934, an 80 ha garden site in the western part of the city was created to help 5000 unemployed families. Market gardens and greenhouse operations were very common in Toronto until the 1930s (Fram, 2009). During WWII, Canadian city Victory Gardens created 200,000 wartime gardens nationally and produced 52 million kg of vegetables (Johnson, 2009). Up until the 1960s, much of the northern part of Toronto was still farmland, but it was gradually converted as population and commercial pressures resulted in redevelopment.

Statistics Canada (2013) reported that 4% of Canadian households grew fruit, vegetables, herbs, or flowers in community gardens, with the total reaching 9% in Quebec. This study did not include the territories and indigenous communities. A 2015 survey of 81 cities found that all had community gardens, and almost half had rooftop gardens. However, in most cities the number of gardens/100,000 residents was low (Soderholm, 2015). Montreal has the most robust community garden program in Canada, with an extensive network of 100 or so community gardens supported by the municipality and community organizations since the 1970s, and providing for about 12,000 residents, with over half identifying as low income (Bhatt & Farah, 2016). The city also boasts an array of commercial and self-provisioning rooftop farms and gardens, including several sites devoted to grapes (http://www.au-lab.ca/recherche/vignes-en-ville/). One estimate has the City producing food across all types of production on 36 ha, amounting to 1000 tonnes of vegetables (Duchemin & McClintock, 2020). Toronto had 100 growing gardens on public properties (Miller, 2017), but not all are directly supported by the municipality. Vancouver has a similar situation. Gardens in some cities were prevented from opening during the first COVID lockdown, but many opened later once municipal officials agreed they were essential. And from 2021 announcements out of Toronto City Hall, it appears municipal officials recognized their error and opened gardens earlier.

A most interesting development is the construction of community gardens and greenhouses in northern Canada. Chen and Natcher (2019) reported that 36 community gardens and 17 greenhouses existed across the North, with 36 located in the Northwest Territories, 10 in the Yukon, 3 in Nunavut, 2 in Labrador, and 2 in Nunavik. Thirty-four (64%) were in communities with less than 1000 residents. A few initiatives also included poultry and a few greenhouses operated hydroponic systems. Some were funded with federal and territorial grants. Many focussed on self-provisioning for members with the excess sold in some cases or donated. Some were structured as individual allotments, others communal. Clearly this is not part of the traditional dietary pattern of many in the North, so in some ways their existence is partial compensation for colonialism and climate change. Although there is the Northern Farming Training Institute in Hay River to help with training needs, food production, whether gardens or farms, is a challenge in many communities that have little tradition of growing food (Spring et al., 2018). It is also sometimes difficult to access land because of soil and climate conditions or government regulations (Schiff & Brunger, 2013).

Urban Aquaculture and Livestock

Because beekeeping and chicken rearing often happen in backyards, it is challenging to develop a clear picture of levels. The 2015 survey of 81 Canadian mid-sized and large cities found that almost 90% permitted urban beekeeping, almost one-fourth permitted urban chicken rearing, and about 7% had urban aquaculture operations present, presumably commercial (Soderholm, 2015). Toronto has about 500 beehives in about 200 yards (Secord, 2019). Some Montreal boroughs, Victoria, Vancouver, Kingston, Kitchener, Niagara Falls, Toronto (pilots), Edmonton, Whitehorse are some of the cities that permit chickens. Anecdotally, many keepers of urban livestock and bees do so for the pleasure and taste of the food, rather than economic needs. There are some reports of restaurants keeping bees for their own use and there may be sales at urban farmers markets.

Urban Foraging

Cities produce significant amounts of food (backyards, municipal properties, rights of way, public trails) that often go unharvested, particularly fruit and nuts, and other kinds of edible landscapes. This has spurred on urban foraging movements, sometimes individual, increasingly through nongovernmental organizations and social purpose enterprises, such as LifeCycles in Victoria (which started in 1994); Not Far From the Tree, in Toronto (2008); Operation Fruit Rescue, in Edmonton (2009); Les Fruits Défendus, in Montreal (2011/2); and Hidden Harvest, Ottawa (2012). Ottawa had 17,000 food-bearing trees in 2012/13, 4000 on city property. Hidden Harvest organizes the collection and splits the harvest, 25% to the nearest food agency (shelter or food bank), with the rest split equally between the homeowner (if private property), the volunteer harvesters, and itself, selling its share to local restaurants and processors to help fund operations. It is structured as a social purpose business (Ballamingie et al., 2019). "In Toronto, there is an estimated one and a half million pounds of fruit growing throughout the city and, yet, less than 17,000 lbs. of fruit was harvested in 2014" (Marshman & Scott, 2019).

Municipal land use planning and regulations are uneven regarding urban foraging and this creates barriers to harvest (McLain et al., 2014). "Some critics view the model of urban harvesting as an inefficient and illegitimate means of producing food and assuring food security in cities that has little to no potential for scalability. In fact, scaling up and out—becoming bigger and more profitable—brings potential liability, exposure, regulation, competition, and criticism" (Ballamingie et al., 2019).

Promoting Self-Provision and Urban Agriculture in Canada: Key Strategies

I use the Efficiency-Substitution-Redesign transition framework to think through solutions (see http://foodpolicyforcanada.info.yorku.ca/solutions). Because self-provisioning is inherently regional in focus, strategies will have to be adapted to local circumstances and not all will be applicable for every setting.

Efficiency Stage

Municipal Official Plan Statements Regarding Food Production and Food System Function

Most municipalities with significant focus on urban food production use a combination of official plans, related plans linked to the Official Plan, bylaws, policies, and programs to bring direction and implementation to their efforts. Some municipal officials argue that activities like community gardens can take place without specific municipal tools, but it appears that those cities with specific interventions are more likely to have more significant activity (cf. Soderholm, 2015). Regarding related plans, a number of municipalities have specific food and agriculture plans, strategies and/or Food Charters, including Victoria, Vancouver, Kamloops, Calgary, Edmonton, Toronto, London, Kawartha Lakes, Montreal, and Halifax (see also MacRae & Donahue, 2013).

Two of the better developed Official Plans that many other municipalities should adapt to their circumstances are those of Victoria, BC and Edmonton (Curran, 2019). Victoria's Official Community Plan has a five-page section on food systems that focuses on food security and sustainability. It contains a wide range of objectives regarding food system planning, access to food, gardening on public and private land, food system infrastructure, and food waste reduction. The other important feature of the plan is a section on adaptive management. Adopted in 2010, the Edmonton Municipal Development Plan contains a five-page section on Food and Urban Agriculture. It also speaks about the need for a local food and a sustainable food system and imagines linked zones of food system activity from the core out to the countryside. It specifically names the need for an Edmonton Food Charter and a city-wide Food and Agriculture Strategy that integrates food production into public and private spaces. The forest management plan of a city is another important area for directive statements. Ottawa's Urban Forest Management Plan includes the recommendations to "Support the planting of edible landscapes and encourage registration of existing edibles on public and private lands" (Ballamingie et al., 2019).

A key challenge of municipal plans is ensuring other instruments bring force to the intentions stated in the plans. Frequently, statements and instruments are misaligned or incomplete. Equally important, it is sometimes ancillary rules and bylaws that prevent implementation of urban food production. An analysis conducted for the City of Hamilton in 2013 showed how a range of rules covering fences, temporary buildings, soil, pesticides, property standards, drainage, signs, and noise can all have restraining impacts on food production (Corey et al., 2013). Ensuring these are consistent with intentions expressed in plans is a critical component of municipal success.

Municipal Bylaws Regarding Food Production and Food System Function

Bylaws are a most concrete way to bring effect to plans, with zoning and animal bylaws the most common subcategories relevant to this discussion. However, cities infrequently specify urban food production as a land use (Soderholm, 2015), with a few positive exceptions.

Zoning

Important examples that can be adapted to other municipalities include:

- Edmonton, AB Bylaw 17,527 permits indoor urban growing in some shopping center and business district zones, outdoor operations in some commercial and community service zones and gardens in residential zones (Curran, 2019).

- City of Nanaimo, BC Zoning Bylaw (no.4500) permits an urban food garden in all zones provided that it does not create odor beyond those on sites without gardens (Part 6).
- Hamilton, ON Zoning bylaw 05–200 allows urban farms (4.26) and community gardens (4.27) in many existing zoning districts, with restrictions on backyards and rooftops.

Important changes to rooftop bylaws and associated programs for food production are required. Although many cities have green roofs, they operate often in a restricted or gray area. Montreal rooftop food producers have done well in a relatively unspecified regulatory environment, but they are seeking greater precision, and this is probably not a viable approach for cities across the country. Admittedly, agriculture on rooftops creates new land use–related challenges to traditional authorities (Curran, 2019), and this suggests clarity is required. Toronto's green roof bylaw focuses on perennial plantings for cooling and water management and would need amendment to incent food production.

Livestock, Fowl, and Beekeeping

Animal bylaws can address urban animals, particularly chickens and bees, in three different ways: explicitly allowed; explicitly prohibited; or not mentioned. The latter category obviously creates a gray zone (Soderholm, 2015). Large animals are usually excluded except in zones specifically identified as agricultural. It is problematic that chickens and bees are often controlled under animal control bylaws with enforcement by animal control units, but this reflects the lack of food production regulatory instruments in most urban areas.

Victoria is interesting because its animal control bylaw forbids roosters and peafowl, but does not name chickens. It appears that from this, city officials have determined that a household can have up to 15 poultry (chickens, ducks, geese). They must be contained on the home property, not in the front yard, or beside a neighbor's property or shared fence. They recommend specific kinds of coops and ask prospective chicken owners to consult animal control before going ahead. Other jurisdictions are more restrictive and specific about numbers/area ratios and aligning numbers with household egg consumption. It seems unlikely that any Canadian municipality needs to ban urban chickens, so all urban municipalities should at least commence pilots to test what rules are most suitable for their circumstance.

For bees, Saskatoon has a very general provision providing for beekeeping with the sole restriction that they be well cared for and don't exhibit defensive behavior. This is interesting in that no distance separation from residences is specified, a provision in many provincial regulations and/or urban bylaws, and one that is often difficult to comply with in denser neighborhoods. Some municipalities have chosen not to enforce their provincial separation requirements as a result and provincial bee officials often take a hands-off approach to the details of urban beekeeping. Calgary has decided that it does not have jurisdiction over beekeeping so does not restrict it unless there is a complaint. Calgary beekeepers are supported by the Calgary and District Beekeepers Association. In most provinces, beekeepers are governed by a provincial Bee Act and must register hives with the provincial apiarist. Vancouver permits bees in community gardens as per its hobby beekeeping bylaw (No.7985), a maximum two hives per parcel less than 10,000 sq. ft.

Municipal Program Supports for Gardening

Montreal is the best model currently because of the engagement of the municipality, degree of collaboration and scale. The city provides the land, equipment, and other materials, does maintenance and repairs on equipment and buildings, provides water, collects garden refuse, and has horticultural animators who visit gardens on a rotating basis and provide advice. The city works with volunteer gardener's associations on enrolment (using a City of Montreal portal) and garden maintenance. Sites are usually institutional, commercial, or industrial. Although city officials have been supportive, community garden organizations have had to keep pushing to enhance the program.

Important program elements to improve success in cities include (see also MacRae et al., 2012):

- Garden (and bees and livestock) resource centers. Many resources exist but not necessarily in coordinated locations linked to municipal garden programs.
- Soil testing and decontamination. Since many gardens are located on abandoned sites, soil quality is a critical issue. Some municipalities have testing and decontamination protocols for community organizations (e.g., Toronto Public Health).
- Garden share (residential, MURB and institutional, indoor, outdoor, rooftops). Whitehorse has proposed doing this in their local food action plan.
- Seed and tool libraries. Seeds and tools are critical for successful gardening. There are many commercial seed companies and an increasing number focus on the urban and ecological marketplace. Many cities already have informal seed exchanges operated by NGOs. Brampton, ON responding to COVID, offered one of the first city-wide programs in Canada to distribute seeds and plants to residents for their own gardens and to share the surplus with local community organizations. Halifax is taking a similar approach. Neighborhood agricultural hubs, as found in Cuba, could provide specialized seeds and seedlings, along with other equipment and services.
- Garden barter markets. Because many municipalities restrict commercial sales from community gardens, barter markets have been established involving garden participants who share between gardens.
- Garden art. Many community and teaching gardens commission murals, sculptures, and environmental art as part of garden beautification. In some cases, municipalities support the artists through public art or more specific programs (cf. programs in Hamilton and Vancouver).
- School gardens.

- Municipal greenhouses. Some municipalities and paragovernmental agencies have greenhouses that are used for starter and bedding plants in municipal gardens. In some cases they are underutilized and all or parts of greenhouses have been integrated into the community garden system.

Improving Composting and Urban Soil Fertility Programs (Adapted from Dirks, 2021)

Soil fertility is of course critical for successful gardening. Excessive application of synthetic fertilizer is problematic and many jurisdictions are encouraging organic fertilizer use as a result. While individual homeowners can make their own compost from kitchen and yard waste, cities and institutional environments experience significant organic waste management challenges, given how much a part of the waste stream is food and yard waste. In Ontario in 2017, for example, organic waste was 32% of the total waste stream, but only 39% was recycled (Ontario Ministry of the Environment and Climate Change, 2017). There remain, however, significant hurdles to making compost from municipal organic waste and when it is feasible, for use in food gardens. Ideally, compost is made once food waste is reduced and where feasible human inedible food is provided to animals.

Divided jurisdiction and unsuitable conceptual frameworks result in suboptimal performance. The federal government is responsible for approving fertilizers and the Fertilizer Act poorly serves ecological nutrient management. Organic waste is regulated primarily under waste management authorities, not those related to soil fertility, both of which are largely provincial, but the provinces have chosen to consider this a waste management issue and rely on municipalities to operationalize their waste management directives.

Municipal approaches to organic waste vary tremendously by province and municipality. In general, municipalities do a better job of single unit residential household organic waste diversion than institutional, commercial, industry, and multi-residential sectors. Even municipalities with significant organic waste diversion do not necessarily prefer organic waste as fertilizer. If the material is composted, it is often done anaerobically in centralized facilities and whether aerobic or anaerobic, typically, the quality of the compost is too low to be used for food production. The challenge then is to put in place programs that make use of organic waste for soil fertility.

Urban Gleaning/Foraging of Fruit and Nut Trees

Urban foraging is a significantly underrepresented agenda for urban planners, and many consider it "illegal" in public spaces, for reasons that are not entirely illegitimate (e.g., fears about biodiversity loss, food safety and contamination, reductions in amenity and aesthetic values), but for which solutions are possible given the right attention (Shackleton et al., 2017).

Cash and in-kind supports are needed for community gleaning organizations. They rely extensively on volunteers, so accessing municipal community grants is a priority. Also, municipalities have equipment and resources that could help with the planting, mapping harvest, and distribution process (Clark & Nicholas, 2013).

Public education on safe and ecologically sensitive gleaning practices, and enhancing access to public sites where gleaning can take place is needed (Ballamingie et al., 2019).

Substitution-Stage

Land Lease Programs for Urban Food Production (Adapted from MacRae et al., 2012)

Although community gardening is happening primarily on public or para-public land, there are often significant opportunities to engage a wider range of institutional actors and private firms, given their significant land holdings, often traditional lawns. Urban farmers also often do not own the land they cultivate. There can be worries, however, about gardening practices, security, water, liability, and who has access to the site. Such matters can be addressed by municipal facilitation that links land with gardeners and provides a lease template.

Setting lease rates at zero is obviously optimal for self-provisioning, but not necessarily realistic in urban areas with high land costs and significant competition for building sites. Landowners may have motives other than maximizing profit for offering land at reduced rate. They may want to see food production next to their residence, to support local residents, to gain environmental benefits (birds, bees, etc.), or to obtain a tax break by making land available to nonprofit or public groups. Cities can create incentives for developers to install food production sites, and disincentives if they fail to do so. Rates are very complex when having to determine "fair market value" for land with a limited set of private uses (e.g., hydroelectric corridors).

Given the complexities in larger municipalities with significant land use pressures, a coordinated leasing system will be needed. Interested landowners, including the municipal government and para-public institutions, could contract with a third-party organization to manage lease arrangements based on templates established by the municipality. The third party would set up the lease arrangements with interested farmers and community groups, taking a small percentage of rents supplemented with revenues from the municipality and foundations to finance its coordination activities. If a third party is managing leases, there is some opportunity for pooled leasing rates for farmers and community organizations, with the leasing agency pooling revenue and then dispersing it differentially to landowners. In this way, commercial activities can subsidize self-provisioning.

Urban Food Forests and Edible Landscapes

With Ottawa being something of an exception (Ballamingie et al., 2019), most cities have not made fruit and nut trees a significant part of their forest management planning and implementation, despite their potential contributions to core forest master plan objectives such as improved air quality, reduced storm-water and heat island effects, wildlife habitat, and canopy coverage (Clark & Nicholas, 2013). Calgary has, however, trialed fruit and nut tree plantings in community gardens, in public parks, as regional orchards, along pedestrian routes, and in urban domestic gardens (Clark & Nicholas, 2013). The scenario analysis developed by Clark and Nicholas of new apple tree plantings with agro-ecological management on vacant land in Burlington, Vermont suggested that urban apples could provide up to one-third of the population's apple consumption.

Similarly, there is typically limited attention to edible landscaping in public spaces. In some municipalities, multifunctional edible landscapes are being designed that include food provisioning, recreation, green and social space, and aesthetic value (Clark & Nicholas, 2013).

Fruit and nut trees can be part of municipal tree selection programs. The Clark and Nicholas analysis of fruit tree and bush suitability in climes similar to southern Canada suggested 30 species could be viable, including low and highbush blueberry, apple, sour cherry, certain grape varieties, some pears, a range of berries grown more in western than eastern Canada, sea buckthorn, and some nuts (hazel, chestnut and filberts).

Restricting Use of Synthetic Chemical Fertilizers in Urban Areas

To be consistent with a Circular Economy approach, urban areas must optimize use of organic materials for urban soil fertility. This means reducing as much as possible synthetic chemical fertilizer applications with high quality compost for use in food production and landscaping. At this stage, many municipalities would be encouraging residents and companies to naturalize their landscapes and reduce the traditional mono-cultural or limited grass lawn. Restrictions would parallel and complement limits imposed through provincial legislation and municipal bylaws on synthetic chemical pesticides. Although cosmetic pesticide use regulation is uneven across the country, significant restrictions are in place but there are exemptions for public health and safety matters, control of invasive plants, sports fields, and golf courses (but they must follow Integrated Pest Management practices).

Most provinces have nutrient management legislation, or nutrient management provisions in other legislation and programs, to regulate nutrient use and application on agricultural lands. The legislation and regulations typically allow for a broadening of what lands are considered under the regulations, what nutrients can be applied, and whether a nutrient management plan is required. In some provinces, this may also require a reconsideration of what defines a farmer, usually a minimum gross sales amount. Amendments to acts and regulations should be introduced to restrict application of synthetic chemical fertilizer in urban landscapes and encourage use of composted material from urban organic sources. As with synthetic chemical pesticides, exemptions could be permitted for landscapes where soil fertility is difficult to manage without fast-acting nutrient release, or where soil conditions have specific nutrient requirements that are not easily available in compost (e.g., sports fields, golf courses, showpiece gardens, and rehabilitation sites). Compost for sale to urban farmers would have to meet the requirements of the federal Fertilizers Act, but compost provided by the municipality to residents for self-provisioning purposes would need only to meet compost quality rules of the province.

Governance Strategies for Urban Food Production (Adapted from MacRae et al., 2012)

In most cities, urban food production is happening in a somewhat uncoordinated fashion. Governmental and nongovernmental actors support projects, different funders offer support, access to land can be a difficult process, and spatial dispersal uneven. In the spirit of creating a more coordinated approach for both self-provisioning and commercial production, MacRae et al. (2012) and Nasr et al. (2010) set out mechanisms in Toronto for governing urban food production. Although these proposals have had some influence on the construction of Toronto's Agricultural Strategy, few of their governance proposals have been implemented.

Complex policy and program environments, such as those related to urban food production, are challenging to govern. A structure and process must aggregate resources for implementing food production across numerous complementary and competing actions and actors. Gaps in jurisdictional and regulatory frameworks can create governance challenges. The range of landowners and building owners and the geographic dispersion of production and distribution further complicate the governance environment.

A multi-stakeholder steering body with staffing from a funded agency is the preferred approach (Nasr et al., 2010). It could be modeled on the Student Nutrition Toronto Partnership, which coordinates the implementation of student nutrition programs for over 125,000 children daily in Toronto schools. In this model, overall governance and policy development would be provided by a steering body representing all the main governmental and nongovernmental actors engaged in the sector. Staffing support would be provided by municipal staff already supporting urban food production. The steering body would have an allocations committee that aggregates resources—land, finances, inputs, expertise—and distributes them to projects based on one funding application. The allocation committee's members would include representatives from government, funding agencies, private donors, and program delivery agencies.

Redesign

Common-Pool Resource Management of Food in Urban Areas

Elinor Ostrom's (2010) principles for common-pool resource management can also be applied to land-based systems, including urban areas, although there is debate about whether urban areas are consistent with Ostrom's principles.

In cities, self-provisioning is connected to the urban commons. Although food producing units are generally small, they are linked to the wider processes of managing a city. Especially for self-provisioning, food is not a private good, and the resources needed to produce/obtain it better lend themselves to collective and public good regimes. Land-based commons are historically thought of as areas of land with shared rights to graze, plant, and forage. "Three essential components are identified from the theory of common-pool resources: (1) a shared resource, (2) a community of resource users and (3) collective governance" and "In the urban context, these components can be recognised as (1) the city and its urban spaces as a shared resource, (2) the city's residents as a community of resource users and (3) residents' participation in the city, through political movements or resource management, as a form of collective governance." (Felstead et al., 2019). In the city, this is made more complex than in classical common pool resources by the competition for space, and the resulting processes of inclusion and exclusion. Harvey (2012) cautions that public space and urban commons are not necessarily the same because creating an urban commons requires a collective effort to manage it.

A key issue then is helping a collectivity to develop trust and the skills to collectively manage, especially given that urban communities are often disconnected, and don't typically share a common past. Neighborhood scale may then be important because it counters some evidence that larger groups are more prone to failure. Divergent levels of income and status can also impede effectiveness (Tedder, 2010). They often unite, instead, around a common threat to a common space, frequently a rebellion against modern forms of enclosure (Felstead et al., 2019). This suggests the need for both training and relations with institutional actors with access to the legal and institutional frameworks of cities to help limit enclosures and coordinate the collectivity (Felstead et al., 2019). Of course, this assumes that public institutions are prepared to fulfill public missions, a concept that has been under siege in the neoliberal era. All this is important to avoid collective action failures (Tedder, 2010).

Foster and Iaione (2017) provide examples of urban projects in 100 "eco-cities" that they feel reflect Ostrom's principles. An emerging body of planning and design literature explores how to "convert" existing urban spaces for food provisioning (cf. Carrot City https://www.torontomu.ca/carrotcity/city.html; Kassie, 2019; de la Salle, 2019). The scale of many remains limited, both spatially and in terms of participation, but they represent initiatives to build upon leading to the redesign stage. There are also modeling studies that forecast some of the possibilities with the right planning and design changes (cf. McClintock & Cooper, 2010; McClintock et al., 2013; Saha & Eckelman, 2017).

In addition to conversion analysis, there are also new city neighborhoods being constructed with food at the center. These new neighborhoods are typically dense, providing a wider range of ecological services, including food (cf. Park 20/20 MasterPlan the Netherlands, https://mcdonoughpartners.com/projects/park-2020-master-plan/; case studies in Donovan et al., 2011).

Conclusion: Financing the Transition

For urban areas, many of the initiatives are about removing obstacles and optimizing existing resources. A prime budget area from which to support urban food production is waste management. Curbside organic pickup and centralized anaerobic digestion systems are hugely expensive. Toronto, for example, spends about $80 million annually on organic collection, distribution, and anaerobic digestion and is under pressure to expand the system to meet waste diversion objectives and a looming Ontario ban on organics in landfill. This figure also does not include organic food waste that ends up in the sewer system (garburators, dumping, etc.) (Dirks, 2021). At a minimum, Toronto could potentially avoid expansion costs, if not reduce operating budgets for waste management and sewage treatment, with significant diversion of organic waste to backyard and community composting and soil fertility initiatives to support food self-provisioning. There may also be public and private savings from restricting urban fertilizer use, applying compost instead. This is already done to some degree in many municipalities but certainly has not been optimized.

References

Agri-Food Analytics Lab. (2020). *Home food gardening during COVID-19*. Dalhousie University.

Alaimo, K., Packnett, E., Miles, R. A., & Kruger, D. J. (2008). Fruit and vegetable intake among urban community gardeners. *Journal of Nutrition Education and Behavior, 40*, 94–101.

Ballamingie, P., Poitevin-DesRivières, C., & Knezevic, I. (2019). Hidden harvest's transformative potential: An example of 'community economy. *Journal of Agriculture, Food Systems, and Community Development, 9*(Suppl. 1), 125–139.

Bhatt, V., & Farah, L. M. (2016). Cultivating Montreal: A brief history of citizens and institutions integrating urban agriculture in the city. *Urban Agriculture and Regional Food Systems, 1*, 1–12.

Boyle, E. (2022). *Mobilize food! Wartime lessons for environmental victory today*. Vancouver; Boyle.

Chen, A., & Natcher, D. (2019). Greening Canada's Arctic food system: Local food procurement strategies for combating food insecurity. *Canadian Food Studies / La Revue Canadienne Des études Sur L'alimentation, 6*(1), 140–154.

Clark, K. H., & Nicholas, K. A. (2013). Introducing urban food forestry: A multifunctional approach to increase food security and provide ecosystem services. *Landscape Ecology, 28*, 1649–1669.

CoDyre, M., Fraser, E., & Landman, K. (2015). How does your garden grow? An empirical evaluation of the costs and potential of urban gardening. *Urban Forestry & Urban Greening, 14*(1), 72–79.

Corey, S., Routley, M., & Wakefield, S. (2013). *Urban agriculture policy, planning, and practice: A report for the City of Hamilton, Ontario*. The City of Hamilton.

Curran, D. (2019). Foodlands protection in Canada: Indigenous and colonial foodscapes in law. In H. McLeod-Kilmurray et al. (Eds.), *Food law and policy in Canada* (pp. 151–174). Thompson Reuters.

de la Salle, J. (2019). Great food streets: Planning and design for urban magnetism in post-agricultural cities. *Urban Design International, 24*(2), 118–128.

Dirks, A. (2021). A community compost exchange manual: Reconnecting municipal organic waste and soil management. MES Major Project, Faculty of Environmental and Urban Change, York University, Toronto.

Donovan, J., Larsen, K., & McWhinnie, J. (2011). Food-sensitive planning and urban design: A conceptual framework for achieving a sustainable and healthy food system. In *Report commissioned by the National Heart Foundation of Australia (Victorian Division), Melbourne*.

Duchemin, E., & McClintock, N. (2020). L'apport Alimentaire de L'agriculture Urbaine Sociale Aux Villes, en Temps de Crise: Le Cas De Montréal". Carnet de Recherche AULAB, AgriUrbain.

Felstead, A., Thwaites, K., & Simpson, J. (2019). A conceptual framework for urban Commoning in shared residential landscapes in the UK. *Sustainability, 11*, 6119.

Foster, S., & Iaione, C. (2017). Ostrom in the City: Design principles for the urban commons. *The Nature of Cities Festival*. https://www.thenatureofcities.com/2017/08/20/ostrom-city-design-principles-urban-commons/

Fram, M. (2009). Greenhouse Toronto, once upon a time. In C. Palassio & A. Wilcox (Eds.), *The Edible City: Toronto's food from farm to fork* (pp. 140–147). Coach House Press.

Harvey, D. (2012). *Rebel cities: From the right to the city to the right to the urban revolution*. Verso.

Johnson, L. (2009). "Revisiting victory: Gardens past, gardens future". In The Edible City: Toronto's food from farm to fork, edited by C. Palassio and A. Wilcox, 58–65. : Coach House Press, 2009.

Kassie, M. (2019). Grow small, think big: Designing a local food system for London, Ontario. *Urban Design International, 24*(2), 142–155.

Litt, J. S., Soobader, M. J., Turbin, M. S., Hale, J. W., Buchenau, M., & Marshall, J. A. (2011). The influence of social involvement, neighborhood aesthetics, and community garden participation on fruit and vegetable consumption. *American Journal of Public Health, 101*(8), 1466–1473.

MacRae, R. J., & Donahue, K. (2013). *Food policy and local governments: Innovation from the bottom up*. Toronto Food Policy Council and Canadian Agri-food Policy Institute.

MacRae, R. J., Nasr, J., Kuhns, J., Baker, L., Christianson, R., Danyluk, M., Snider, A., Gallant, E., Kaill-Vinish, P., Michalak, M., Oswald, J., Patel, S., & Wekerle, G. (2012). Could Toronto provide 10% of its fresh vegetable requirements from within its own boundary? Part II: Policy supports and programme design. *Journal of Agriculture, Food Systems and Community Development, 2*(2), 147–169.

Marshman, J., & Scott, S. (2019). Gleaning in the 21st century: Urban food recovery and community food security in Ontario, Canada. *Canadian Food Studies / La Revue Canadienne Des études Sur L'alimentation, 6*(1), 100–119.

McClintock, N., & Cooper, J. (2010). Cultivating the commons: An assessment of the potential for urban agriculture on Oakland's Public Land. http://www.urbanfood.org.

McClintock, N., Cooper, J., & Khandeshi, S. (2013). Assessing the potential contribution of vacant land to vegetable production and consumption in Oakland, California. *Landscape and Urban Planning, 111*, 46–58.

McLain, R. J., Hurley, P. T., Emery, M. R., & Poe, M. R. (2014). Gathering "wild" food in the city: Rethinking the role of foraging in urban ecosystem planning and management. *Local Environment: The International Journal of Justice and Sustainability, 19*(2), 220–240.

Miller, S. (2017). *City region food systems project of Toronto and the greater Golden horseshoe.* RUAF, Centre for Sustainable Food Systems and FAO.

Nasr, J., MacRae, R., & Kuhns, J. (2010). *Scaling up urban agriculture in Toronto: Building the infrastructure*. Metcalf Foundation.

Ontario Ministry of the Environment and Climate Change. (2017). *Discussion paper: Addressing food and organic waste in Ontario.* Ontario Ministry of the Environment and Climate Change.

Ostrom, E. (2010). Beyond markets and states: Polycentric governance of complex economic systems. *The American Economic Review, 100*(3), 641–672.

Patel, R. (2012). *Stuffed and starved: The hidden battle for the world food system.* Melville House Pub.

Saha, M., & Eckelman, M. J. (2017). Growing fresh fruits and vegetables in an urban landscape: A geospatial assessment of ground level and rooftop urban agriculture potential in Boston, USA. *Landscape and Urban Planning, 165*, 130–141.

Schiff, R., & Brunger, F. (2013). Northern food networks: Building collaborative efforts for food security in remote Canadian aboriginal communities. *Journal of Agriculture, Food Systems, and Community Development, 3*(3), 31–45.

Secord, P. (2019). Can cities help feed cities: Can 10% of Toronto's honey available for consumption be met with urban apiaries?" MES Major Project, Faculty of Environmental and Urban Change, York University, Toronto.

Shackleton, C., Hurley, P. T., Dahlberg, A. C., Emery, M. R., & Nagendra, H. (2017). Urban foraging: A ubiquitous human practice overlooked by urban planners, policy, and research. *Sustainability, 9*, 1884.

Soderholm, D. (2015). *Planning for urban agriculture in Canadian cities.* MA Planning, U. Waterloo.

Spring, A., Carter, B., & Blay-Palmer, A. (2018). Climate change, community capitals, and food security: Building a more sustainable food system in a northern Canadian boreal community. *Canadian Food Studies / La Revue Canadienne Des études Sur L'alimentation, 5*(2), 111–141.

Statistics Canada (2013). Canadians and Nature: Homegrown fruits, herbs, vegetables and flowers, 2013. Catalogue no. 16–508-X, Statistics Canada, Ottawa.

Tedder, S. J. (2010). Common pool resources and state intervention: Why, when and how. PhD Diss., UBC, Vancouver.

Chapter 6
Economic and Commercial Urban Farming

Nabil Kamal

Abstract The adoption of new technologies in vertical farming (VF) businesses is now quite common and motivated by a complex web of factors. It is more likely to expand given the high yield, out of season crops, and cost-effective savings of water and land. The impacts of this adoption are expected to disrupt our daily life as countries, organizations, farmers, and consumers. This novel technology addresses the issues of "Food Security," "Supply Chain," and "Food Safety," all at once. VF technology is progressing into becoming a fully mature and reliable business. Developments in the fields of renewable energy, artificial lighting, robotics, and AI will contribute to this end and expand and deepen its uses and effectiveness.

Keywords Urban agriculture · Vertical farming

Introduction

When the early electric vehicles (EV) models were introduced to the market, a common narrative was in sway that this technology was only for the elites, that everyone cannot afford it, that it is only meant for the sports racing industry, and that it would never match the reliability, feasibility, or quality of the traditional internal combustion engines (ICEs). This EV technology was so disruptive that it has changed the existing perception of the car industry within a very short time, it is now seen as a technology accessible by the ordinary people and thus it has significantly eroded the social inequities that were first anticipated. Similarly, the vertical farming (VF) technology was first thought of as an exclusive technique for the high-end expensive elitist food market. However, the high-tech farming technologies, including the VF, are adapting different techniques which have gained momentum to constitute a real disruption to the food industry.

N. Kamal (✉)
PerTech Industries Canada Inc., Burlington, ON, Canada

A. A. Kubursi et al. (eds.), *Urban Food Security in a Crisis Prone World*,
https://doi.org/10.1007/978-3-031-89440-4_6

So, how would this VF technology affect our daily life as countries, organizations, farmers, and consumers? Taking into consideration the global serious challenges that humanity is currently facing in regard to food security, food safety, food quality, global warming, environmental challenges, water scarcity…. Would High-Tech Farms be a viable solution to many of these problems and challenges?

First, let's understand this technology and its present standing; in the last decade or so, those futuristic High-Tech Farms have spread around like wild mushrooms, they are integrating different technologies ranging from hydroponic farming with simple controls all the way to the very sophisticated fully artificial intelligence (AI) controlled systems that energize special farming robots, totally encapsulated in a strictly controlled climate, using renewable energy sources, LED artificial lighting, Aquaponics, and Aeroponics farming.

The VF is a relatively new concept for engineering food production; it is simply about controlling all the inputs and required resources so as to get the desired optimized output, i.e., food. The VF concept is an attempt to address some of the imminent global challenges, such as (Just Vertical, 2022; The Grower Inc. 2021; TODAY, 2017)

- Population growth that is leading to the ever-increasing demand for food (10 Billion by 2050).
- The increase in the standard of living leading also to a demand increase for food (The United Nations estimates that on account of the anticipated global population growth, global food demand will rise 70% by 2050).
- The increased urbanization trends of the population where more than 60% of the population will live in urban areas.
- The disappearance of billions of tons of topsoil every year (About 25 Billion Tons of lost soil per annum).
- The decrease in global agricultural yields due to global warming-related issues, and also due to the drastic change in climate patterns.
- The increase in the emissions of greenhouse gases (25% by 2030).
- The long-distance trips between the producing farms and the markets (some fresh produce travels 4000+ kms to reach markets), and.
- The high percentage of food wastage due to diseases, mishandling, and changing consumer tastes, among other reason.

Coping with the Stresses in Agriculture and Food Systems

There are many stresses facing our planet. Agriculture is the cause of some 70% of the global freshwater contamination and about 70% of the fresh water usage goes to agriculture. One-third of the arable land has been degraded in the last 40 years. Planet Earth is overloaded with stresses that it cannot cope with; science has to develop new technics for modern farming.

Over the past few years, vertical farms have spread all over the world, including in Canada, Japan, Singapore, Panama, the UK, UAE, and around many areas in the USA. The prominent players in the existing high-tech VF industry report some outstanding and very promising results:

- In comparison to a traditional farm that produces 8.2 heads of lettuce per m^2 on average, the VF can produce up to 1640 heads per m^2. That is due to the high-efficiency year-round operation with consecutive number of harvests in up to 16 levels' structures. The VF climate-controlled systems reduce the growing cycle (harvests) by 50%, produce can be grown all year round, with a much smaller impact on the environment.
  The VF root misting system allows the use of 95% less water than a regular field farm (Anon, 2016).
- The VF, being totally enclosed and operated under a completely controlled environment with strict hygienic procedures, does not require any pesticides or herbicides.
- Positive environmental impact: instead of soil, plants are grown in reusable organic media.
- As the VF uses specialized LED lighting, it was shown that changing the light recipe using the control system would alter the physiological response of the plant. Thus, different light combinations do manipulate size, shape, texture, and taste. Furthermore, it was also shown that the produce could be enhanced by making them more nutritious as it is feasible to increase content such as calcium and phosphorus and various vitamins by as much as 50% just by changing light recipes.
  (CBC News, 2018).

The Pandemic Effects

The COVID-19 pandemic has been a great shock to individuals, governments, businesses, and all sorts of organizations. It was clear evidence of how fragile and how interdependent the world has become.

Similar to "global warming," some buzz words that were alien to the public a few years ago had become a fact of life that all of us have to deal with: "Supply Chains" and "Food Security" became a major concern on different levels. The Pandemic has changed our personal and national priorities; Food security has become on the top of the list of priorities for governments worldwide. We are witnessing a substantial increase in the demand for food, grain storage, and food processing systems; different countries are reassessing their national Food security and Food safety strategies; and multiple projects are being developed accordingly.

Additionally, one major hurdle during the pandemic was the complete halt of the cross-border travel of the seasonal workforce, which led to major difficulties in all traditional farming activities. In numerous cases, fruits were left out in the fields uncollected because of labor shortages.

Can Vertical Farming Technology Solve the Problem?

The VF technology is most probably the only feasible solution currently available that can address the issues of "Food Security," "Supply Chain," and "Food Safety," all at once. It represents a credible recipe to deal with these serious problems. (PBS News, 2017):

1. The substitution of local production of fresh food for importing it from other countries thousands of miles away.
2. Being an urban farming concept, the VF will in the future spread all over the urban centers, which will have a great and noticeable positive impact on food supply chains, and will simplify the currently sophisticated, problematic, costly, and environmentally unfriendly fresh food supply chain. As mentioned before, a good percentage of fresh food is currently traveling for thousands of miles and crossing multiple international borders to reach the end consumer.
3. Being operated under a completely closed environment, with strict control and traceability of all inputs, and with the minimum interference of human workers, the VF affords an unprecedented very high level of food safety.
4. Producing locally fresh food creates jobs, saves foreign exchange, transportation, and storage costs. The avoided costs alone constitute significant resources that can be re-spent on other commodities and services.

As such, governments in many countries are encouraging and supporting the public use of this new farming concept leading to greater acceptance of the public of this system.

In some countries, people are now increasingly witnessing the usage of VF units at their everyday grocery shops. Equally, there is evidence that shows that people do appreciate collecting their own fresh food from the grocery shop where VF stands are being advertised and that their hands will be the first ones touching their food. In the not-so-far away future, it's foreseen that similar VF stands will be a part of many household systems, affording customized and on-demand fresh produce.

VF Fresh Food Production

For the time being, the VF technology has proven to be feasible for the Leafy Greens Vegetables, which constitute only a small portion of the complete family diet. The major issues about the feasibility of this system are the large capital investment requirements for such sophisticated systems and the high running cost of energy that is required for artificial lighting.

The rapid development of LED artificial lighting technology and the increase in renewable energy production and storage are significantly and positively affecting the VF feasibility.

In General, It Is Foreseen That the VF Vertical farming (VF) Technology Will Undergo Different Stages (Anon, 2021).[1]

The first stage is about the production of different types of leafy green vegetables. This stage is widely accepted to be fairly mature by now where there are hundreds of successful enterprises that are in the businesses of producing such products in different countries. There is still significant room for further development to the standardization and the optimization of all production parameters. Rapid outstanding developments are currently undergoing full automation of such systems including the implementation of different AI systems; this will accelerate the accumulated experience and will lead to the optimization of all the process parameters.

The second stage is when the common vegetables become feasible to be produced in the VF technology. This stage is about vegetables such as tomato, cucumber, zucchini, eggplant, bell peppers, cauliflower, and also underground vegetables such as potatoes.

The VF systems are capable of producing such veggies in principle, however, as the fruiting process of such vegetables requires considerable amounts of energy; this VF technology stage is seen to be unfeasible for the time being.

There have been some serious attempts to get into this stage. The new above-mentioned innovations and techniques are rapidly contributing to the development of this second stage. Some researchers have estimated a time span of 10–20 years for this stage to forge itself as a feasible and reliable business. In my opinion, the challenges posed by the pandemic will shorten this time span and will increase the likelihood of introducing these products in the near future.

Once this stage is achieved, a direct impact will be realized by all of us. This will be a major disruption to the farming industry, consumer habits, and will lead to increased urbanizing trends for the traditional farming communities, which will pose some new challenges at the time.

This third stage is about the mass production of stable foods, pulses, grains, and even fruits' growing trees. This stage will require very advanced technologies and substantial accumulated knowledge. It will probably take a few years before we get into this stage.

It is worth mentioning here that the VF systems are also currently used for the successful production of fresh fodders for the cattle and poultry industry. These are very similar to the first stage mentioned above. The use of fresh fodders is gaining momentum with so many farmers opting to target the top market segments.

[1] https://www.youtube.com/watch?v=x5QTyxZW040&t=122s&ab_channel=ExaCognition

Farming: Open-Loop/Closed-Loop Control Systems

From an engineering point of view, traditional farming is looked at as being an open-loop control system, that is (according to Wikipedia) a control action from the controller is independent of the "process output," which is the process variable that is being controlled.

In other words, in an open farming field, it is next to impossible to get a timely impact of changing any of the parameters on the general process performance as there are quite a large number of parameters related to the seeds, water, soil, air quality, seasonal effects, environment, climate, diseases, rodents, etc.

The effect of changing any input will require a long time, probably one complete harvest production cycle or even more, to be realized. And even then, there will be a need for guessing and multiple estimations in the process. That's why farming has always relied on the expertise of the farmers that are passed from one generation to another.

Being operated in a completely controlled and closed environment, the VF technology affords the isolation of different variables and will provide for a much easier and more efficient cause-effect analysis. This will lead to even more understanding of the farming process itself leading to greater efficiency and to the proper usage of resources.

The new research about detecting and interpreting plant reactions to different stress modes in the form of plant noises will certainly be integrated in the future in the VF technology. In this respect, we can easily say that farming processes are on the verge of being transformed from open-loop to closed-loop control systems engineering-wise, which will have a great impact on efficiency, feasibility, and on the environmental impact of the process.

The Future of the Farming Industry

In the broad sense of the term, "Farming Industry" means the farming business; however, it is my opinion that the VF technologies and new related innovations are transforming the nature of the farming process from being a mere agricultural to an industrial process that is fully developed and completely automated.

Let us remember here that a milestone in our history was when early humans created agricultural communities, which were the nucleus of urban centers, social interactions, and cultural development leading to the present sophisticated civilizations. Another humankind milestone was related to the industrialization and mechanization of production processes leading to the further development of societies and civilizations. It seems as if humankind is about to witness a convergence of the two previous milestones into a new one that combines both agriculture and industry, where humankind will have full control over its food resources.

Conclusions and Recommendations: The Future Outlook of VF

As space and resources become more and more scarce, farmers are finding innovative solutions to so many problems. Farming is getting turned on its head including going away from the fields and taking it indoors, stacking the plants vertically, and growing their crops in the absence of sunlight.

It goes without a doubt that the VF technology will progress into becoming a fully mature and reliable business. Developments in the fields of renewable energy, artificial lighting, robotics, and AI will contribute to that end.

Governments, specialized businesses, and concerned research institutes are strongly recommended to invest in the future of VF farming and to include this technology in their strategic plans. There is also room for small farmers, entrepreneurs, and start-ups to get into this line of business.

Being involved in this industry and other similar high-tech ones, I can easily say that just like all other disruptive innovations, their fruits will be harvested by the first entrants. It seems that the VF technology fruits are about to ripen.

References

Anon (2016). This farm of the future uses no soil and 95% Less Water. https://www.youtube.com/watch?v=-_tvJtUHnmU&t=42s&ab_channel=Stories

Anon (2021). Exa cognition: What's the state of vertical farming in 2021? https://www.youtube.com/watch?v=x5QTyxZW040&t=122s&ab_channel=ExaCognition

CBC News (2018). Hydroponic farming looks to offer food stability across Canada. https://www.youtube.com/watch?v=GjCDQgjRcMo&ab_channel=CBCNews%3ATheNational

Just Vertical (2022). Tackling food insecurity through urban farming technology. https://justvertical.com/blogs/all-blogs/tackling-food-insecurity-through-urban-farming-technology

PBS News (2017). Could indoor farming help address food shortages? https://www.youtube.com/watch?v=7J9f59usLfI&ab_channel=PBSNewsHour

The Grower Inc (2021). Sustainably grow local. Fresh food year-round. https://www.thegrowcer.ca/?gclid=CjwKCAjw6MKXBhA5EiwANWLODME4J4e5Qfp6yIKLXXwJZViaw1snu7PTer6J6riWl0PDO2k_-zFqThoClGsQAvD_BwE

TODAY (2017). Vertical farming is revolutionizing how food is grown in America. https://www.youtube.com/watch?v=Mh_9Dn0Y6bQ&ab_channel=TODAY

Chapter 7
The Role of Livestock Production in Urban Food Systems: Urbanization and Urban Consumer Demand Trends for Livestock Products

Emma C. Stephens, Karen Landman, Karen Schwartzkopf-Genswein, Daryoush Alipour, Ali Kiani, Monika A. Gorzelak, Roland Kröbel, Elham Rahmani, and Mohammad Reza Marami Milani

Abstract By 2050, two-thirds of the world's population is expected to live in urban areas. As a result, global agri-food systems will face increasing pressures to supply growing urban consumers' diets. Greater attention is being given to the potential for urban-based agricultural production to fulfill some of these demands. However, within these discussions, relatively little focus has been on the unique role of urban livestock agriculture and how it might also contribute to feeding the world's cities. In this chapter, we describe three distinct urban livestock initiatives in Canada, Iran, and Germany and compare their features and objectives. We then draw upon these comparisons to identify important characteristics that are needed to support urban

E. C. Stephens (✉)
Agriculture and Agri-Food Canada (AAFC), Northwestern Regional Office, Canada Place, Edmonton, AB, Canada
e-mail: emma.stephens@agr.gc.ca

K. Landman
The School of Environmental Design and Rural Development (SEDRD), University of Guelph, Guelph, ON, Canada
e-mail: klandman@uoguelph.ca

K. Schwartzkopf-Genswein · M. A. Gorzelak · R. Kröbel · E. Rahmani
AAFC, Lethbridge Research Centre, Lethbridge, AB, Canada
e-mail: karen.genswein@agr.gc.ca; monika.gorzelak@agr.gc.ca; roland.kroebel@agr.gc.ca; Elham.rahmani@agr.gc.ca

D. Alipour
Bu-Ali Sina University, Hamedan, Iran
e-mail: alipourd@basu.ac.ir

A. Kiani
Justus-Liebig University of Gießen, Giessen, Germany

M. R. M. Milani
Freelance Food Science Researcher and Data Analyst, Lethbridge, AB, Canada

A. A. Kubursi et al. (eds.), *Urban Food Security in a Crisis Prone World*,
https://doi.org/10.1007/978-3-031-89440-4_7

livestock systems and that can enable their development and contribution to urban food systems overall.

Keywords Agriculture · Livestock · Urban food production

Urbanization and Livestock Production Overview

More than half of the world's population already lives in urban areas and by 2050 this will grow to more than two-thirds (Ritchie & Roser, 2018). Urbanization is increasing in all regions of the world, at rates between 0.3% and 1.3% per year, with cities in Africa and Asia at the top end of that range and predicted to grow the fastest until 2050.[1] Cities are net consumers of food (Schreiber et al., 2021) and the vast majority of urban consumers are not engaged in producing any food for their own consumption, despite increasing innovation in urban agriculture and food production (O'Sullivan et al., 2019). This increased population concentration within cities often means out-migration from rural areas, with falling employment in the agricultural sector and structural transformation away from agriculture and into manufacturing and services (Michaels et al., 2012). These two big shifts have led to the development of a complex food value-chain infrastructure to supply food from rural areas (both local and global) to flow into these growing cities (de Bruin et al., 2021; Schreiber et al., 2021).

When considering how urban agriculture can help meet the food requirements of cities, livestock agriculture poses a unique challenge. Meat and dairy consumption demand is driven by both population increases as well as increasing income levels to a greater degree than other agricultural commodities[2] and these forces are particularly concentrated within the world's growing urban populations. Thus, it is anticipated that global demand for livestock agricultural products (and related feed crops) will increase in the coming decades as a share of global diets (Bijl et al., 2017), with the largest increases expected in developing nations where incomes and populations are growing the fastest.[3] It is apparent that much of this increased demand will be concentrated in urban areas within these countries.

If livestock product demand is expected to grow, particularly for urban consumers, this raises questions about the pressure this may exert on the various livestock value chains that meet the demand of the world's cities currently and how urban livestock production might respond to these pressures. Historically, livestock value chains tended to be relatively short in comparison to crop agriculture, due to the lack of cost-effective transportation, cold chain technology, and high enough

[1] Population trends taken from: United Nations (UN). D. E. S. A. (2018). Revision of world urbanization prospects. New York: United Nations Department of Economic and Social Affairs.

[2] OECD-FAO (2020). OECD-FAO Agricultural Outlook. OECD-FAO Agricultural Outlook 2020-2029. https://doi.org/10.1787/1112c23b-en

[3] FAO (2017). The future of food and agriculture–Trends and challenges. Annual Report, 296.

economic returns—meaning livestock production often had to be near consumers and market hubs (Capello, 2021). Thus, in the past, urban livestock production was a common feature of life in cities around the world.

Livestock value chains today tend to be much longer and animal production for the most part has moved out of cities with the advent of improved transportation technology, and has even extended over greater distances on an international scale. For example, Fig. 7.1 (Bijl et al., 2018) highlights a great degree of intercontinental trade in animal products. Urban areas today are therefore supplied by much longer regional and global livestock value chains, with the greatest pressure on these value chains coming from the fastest growing urban areas. Urbanization is an important driver of agricultural and food system change worldwide (Seto & Ramankutty, 2016; de Bruin et al., 2021), including livestock production and trade.

Renewed attention is being paid to the potential of urban agriculture to supply increasing food demands from cities, including urban production of animal products. Urban livestock keeping has been practiced for millennia all over the world,[4] and fulfills a variety of needs, from increased food security, to satisfaction of consumer demands for local animal products that are produced with specific attributes, such as organic or heritage food designations, a desire to diversify urban economic activity or repurposing vacant urban land and spaces (e.g., https://landoverlandings.com/). In comparison to conventional livestock agricultural production, urban livestock has received relatively little attention from the agricultural research

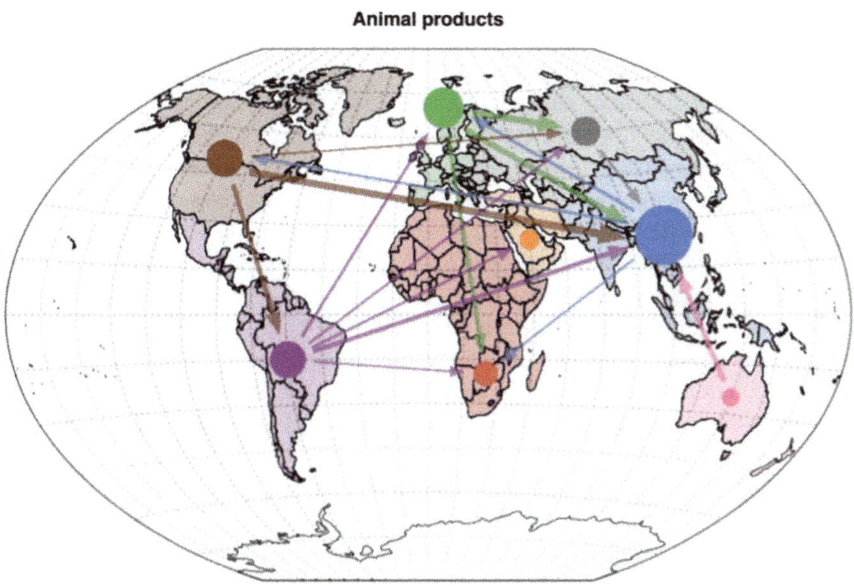

Fig. 7.1 Primary continental trade flows of animal products (meat, dairy) (Bijl et al., 2018)

[4] FAO, Schiere, H., & van der Hoek, R. (2001). Livestock keeping in urban areas: A review of traditional technologies based on literature and field experiences.

community (O'Sullivan et al., 2019). However, if it is to continue to play an active and significant role in provisioning urban food demand, the urban agricultural sector must have a substantial increase in productivity in trajectory similar to conventional livestock production systems, but under restrictions such as zoning and limited available land, nuisance ordinances, and much more (https://ucanr.edu/sites/UrbanAg/). With examples from Canada, Iran, and Germany, we intend, in this chapter, to highlight these different forces at work by showcasing urban livestock agricultural production in each area and compare and contrast these efforts and what they suggest about urban livestock systems in the future in the context of growing urbanization worldwide. We will conclude with information on possible paths forward for continued development of urban livestock production systems with a view to maximizing their potential contribution to urban food needs.

Examples of Urban Livestock Systems in Canada, Iran, and Germany

The concept of animal production within cities is not new. European explorers sent to colonize North America and other continents in the late 1500s had to follow laws (i.e., 1573 Law of Indices) regarding colony (village or city) planning, which included the incorporation of common areas for pasturing livestock so that food sources could be raised close to their inhabitants (Brinkley & Kingsley, 2018). Some examples of this include Boston's Commons and Philadelphia's four main green squares for pasturing (Brinkley & Kingsley, 2018). During the nineteenth century, domestic animals (including cattle, horses, pigs, and poultry) played an important role in the development of industrial cities and significantly contributed to their economic growth (Kheraj, 2015). Urban animal production (UAP) also represented an efficient way of producing cheap food since livestock could be fed organic wastes (crop byproducts and spoiled or excess produce) inconsumable by humans, thus also controlling pests and unsanitary conditions within urban centers (Brinkley & Kingsley, 2018; Kheraj, 2015).

Over the course of the nineteenth century, municipal governments developed increasingly restrictive bylaws, pound systems to contain stray animals, and public health regulations to control urban animal production (Kheraj, 2015). These rules emerged as cities grew and the population density of both humans and animals increased, contributing to overcrowding and pollution. Under these conditions, some city dwellers came to view UAP as both a nuisance and hazard while others still relied on it for food and money (Brinkley & Kingsley, 2018; Kheraj, 2015). Another factor contributing to laws restricting or prohibiting UAP was the industrial revolution, characterized by more sophisticated road infrastructures and vehicles, farm equipment and railways, meaning that goods could be transported farther in greater amounts at a lower price. These technological advances reduced the need for food to be raised within the city and facilitated the movement of animal agriculture to rural areas where land was available and intensive production could flourish based on economies of scale as well as increased processing ability such that meat, milk, and egg products could be transported back into the urban center for sale by retailers (Brinkley & Kingsley, 2018). At the same time, there was a growing concern that raising animals in urban centers was dangerous, dirty, and could result in increased human disease. These factors contributed to significant changes in land use patterns, and real estate speculation further increased the rural/urban divide and thus the increase in rural livestock production.

Urban livestock production systems today have arisen to resolve a wide variety of agri-food system needs and concerns, drawing from history but also moving in new directions. There is thus significant variation in the history and development of these systems across urban areas around the world.

City of Guelph (Canada): Facilitating Backyard Chickens and Poultry Raising

The City of Guelph has long allowed for chickens within the city boundaries, but the recent increased interest in urban agriculture has resulted in the City updating their 1985 bylaw as it relates to chickens and other animals. Currently, poultry, ducks, geese, and pigeons are regulated under the City's 2016 Animal Control Bylaw. Residents of the City who have chickens are required, under a 2018 amendment to the bylaw, to register the number of chickens they have, and to indicate whether the resident is a property owner or a tenant. No roosters are allowed within the City limits, to avoid noise complaints, and all feed is to be stored in an animal-proof secured container. The bylaw states that up to ten hens are allowed but that they must be kept in a pen, or coop, with a solid floor that is bedded with material that absorbs chicken manure and that is cleaned and disinfected regularly; the owner must sanitarily dispose of the manure.[5] It is advised that there should be a plan in place for caring for the chickens through the winter months as well.

Other urban municipalities have drawn on the City of Guelph's history and experience with urban chickens, as city councilors and staff respond to citizens' requests to be allowed to keep hens. In early 2018, the City of Toronto launched the UrbanHens TO Pilot Program; the City provides information on what neighborhoods are allowed to participate in the program, registration of the hens, details on chicken coop construction and location, waste disposal, etc.[6] The nearby Region of Peel[7] developed an Urban Agriculture Discussion Paper in order to support

[5] City of Guelph (2016). Bylaw Number (2016)-20,122. https://guelph.ca/wp-content/uploads/animalcontrolbylaw.pdf

[6] City of Toronto (2021). https://www.toronto.ca/community-people/animals-pets/pets-in-the-city/backyard-hens/

[7] Region of Peel. (2019). Urban Agriculture Peel 2041 Discussion Paper. https://www.peelregion.ca/officialplan/review/pdf/urban-agriculture-discussion-paper.pdf

lower-tier municipalities in creating urban agriculture strategies; the Region defines urban agriculture simply as the growing of crops within an urban settlement area. The Paper recommends the upcoming Regional Official Plan Review as an opportunity to strengthen urban agriculture policy to support local municipalities. One of those local municipalities, the City of Mississauga,[8] is currently running a public engagement campaign to determine interest in a city-wide Urban Hen Program.

With the increased interest in urban hens, and with concerns for human health, the City of Guelph reached out to the Poultry Industry Council (PIC) to host Raising Backyard Chickens workshops and webinars. Along with the Ontario Ministry of Agriculture Food and Rural Affairs Poultry Specialist, the PIC provides webinars and workshops to urban chicken owners on housing, chick care, egg quality, feed and nutrition, and vermin and disease management.[9] Participants are also provided with information on feed and equipment suppliers, and with names of local veterinarians who have a food animal license.

The University of Guelph, situated within the City of Guelph, is also assisting with education and information on backyard chickens.[10] Poultry experts at the Ontario Veterinary College (OVC), the University's Animal Health Laboratory (AHL) and the Ontario Animal Health Network have provided workshops for non-poultry veterinarians on small flock medicine, regulations regarding backyard poultry, how to complete a postmortem on chickens and collect samples as necessary, and how to conduct humane euthanasia if needed.[11] Researchers at the OVC and AHL have conducted research on the antimicrobial susceptibility of enteric bacterial in small poultry flocks in Ontario, which highlighted the necessity of careful use of antimicrobials in order to limit multidrug resistant bacteria (Varga et al., 2019). There is concern that pathogens found in the chickens could cause gastrointestinal illness in humans; it is recommended that frequent hand-washing be practiced to reduce transmission between birds and people (Varga in the Crest 2019). Dr. Scott Weese, Director of the Centre for Public Health and Zoonosis at the University of Guelph, is working with local public health units to learn more from backyard chicken owners.

The City of Guelph responded to the increased interest in urban chickens, largely resulting from the increased interest in growing one's own food, by regulating the number of hens and the maintenance of the health of these animals so as to reduce animal welfare concerns from neighbors, and by not allowing roosters so as to reduce noise complaints. As well, the City has partnered with expertise at the local university and with the provincial ministry that is responsible for agricultural animal health, in order to raise the education level of keepers of hens. These efforts have greatly increased the acceptability of urban hens within the city boundaries.

---

[8] City of Mississauga (2021). https://yoursay.mississauga.ca/urban-agriculture-strategy

[9] PIC. (2021). https://www.poultryindustrycouncil.ca/education/raising-backyard-chickens

[10] The Crest, No. 27. (2019). A Look at Backyard Chickens. University of Guelph. https://ovc.uoguelph.ca/sites/default/files/users/k.mantel/files/CREST_SF2019_webversion_a.pdf

[11] OVC. (2021). https://ovc.uoguelph.ca/news/education-research-and-outreach-top-mind-backyard-chickens

City of Golbaf (Iran): Building a Resistive Economy through Urban Livestock Production

The population of Iran has doubled from 42.4 million in 1982 to 88.4 million in 2025 and based on the current growth rate, it is projected to reach 94.8 million in 2035 and over 100 million in 2050.[12] Due to rural-urban migration, this increase in population is mostly in the urban areas and represents an expansion of the urban consumer population, rather than in the nomadic or rural agricultural producer population. Currently, 75.8% of the population lives in urban areas and 24.2% live in rural villages.[13]

For the most part, urban development policies in Iran have focused mainly on non-agricultural sectors, instead of investing in strategies to promote urban agriculture (Pilehvar, 2021). Furthermore, in the livestock sector, policies during recent years were based on moving farms still in urban areas to the areas out of the cities, and urban livestock production has not been much considered by policymakers. On top of this, as the value of land for economic development increases, the returns from construction and other nonagricultural investment is much higher than the income from the sale of farming products, leading to further conversion of agricultural lands around cities to urban uses to support these growing cities (Jamini et al., 2017).

Nevertheless, as in many other countries, many inhabitants in urban areas are interested in backyard livestock production, such as raising poultry to access fresh eggs or just as a popular pastime. However, typical issues like disagreeable odors, noise, waste disposal, pests, and concerns about the spread of disease are barriers to keeping chickens in an urban area. According to current regulations in Iran, poultry farms, even small-scale poultry production in urban areas, are not allowed while in rural areas, keeping livestock and poultry does not require any permits. Breeding some poultry in poultry research facilities and teaching units of some universities can be considered as only recognized examples of urban poultry production, but it has not yet been considered as a strategy for urban farming or regional sustainable development (Mohammadi & Soleimani, 2014).

More recently, there has been a political push for economic resistance and sustainability for the agricultural sector in Iran. It is based on the use of country and local resources and the increased participation of individuals in prioritized economic activities. One such effort has been to pilot urban goat and sheep farming across three Iranian provinces of Kerman, Western Azerbaijan, and Lorestan, to increase participation in urban livestock production.

One city in the urban goat farming pilot is Golbaf, in Kerman province (Fig. 7.2). Golbaf has a long history as a production center of carpet weaving and livestock

[12] Statistical Center of Iran. (2021a). Population and Housing Census, Statistic Detailed Results of Iran. https://www.amar.org.ir/english/Population-and-Housing-Censuses

[13] Statistical Center of Iran (2021b). Political and Country Divisions Census, Statistic Detailed Results of Iran. https://www.amar.org.ir/english.

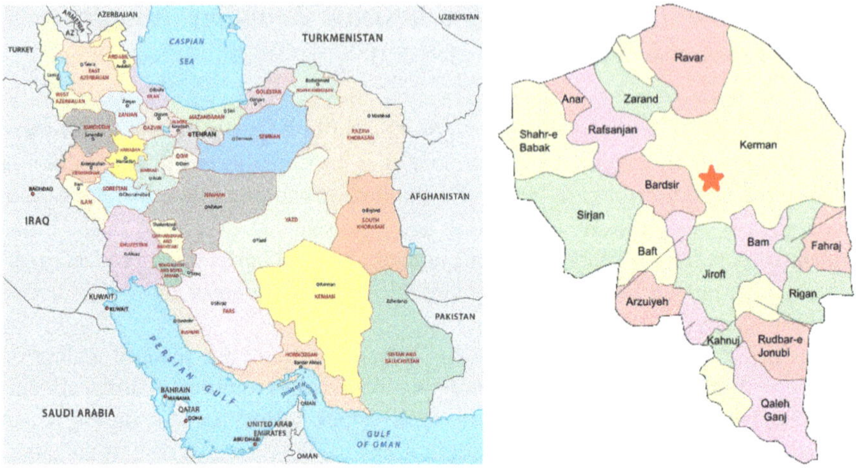

Fig. 7.2 Kerman province in Iran (left map) and Golbaf district in southwest of Kerman county in Kerman province (the red star in right map). Source: https://www.worldatlas.com/maps/iran#admin1Section, https://en.wikipedia.org/wiki/Kerman_County#/media/File:Kerman.svg

farming. In 1981, after an earthquake in Golbaf, in addition to falling demand in global markets for the carpet industry and fewer employment opportunities, people began to migrate from the city and the new program aims to reverse these trends.

The program brings together different supporting resources to facilitate urban livestock enterprise. The Resalat Islamic Social Bank provides finance to resolve various social issues in Iran, such as migration and unemployment (Mirjalili et al., 2020). The Social Development Network division within the bank supports microfinance for small groups of borrowers, with the objective of increasing access to finance for these groups for economic development opportunities. In addition to social loans and microfinance, Resalat Bank provides financial guidance and training to support economic development projects, and the bank has applied these resources to support the urban goat farming initiative in Golbaf.

In this project, a locally trained mentor with an (MSc) degree in animal nutrition is responsible for providing guidance for participating families. Currently, 2843 out of 4162 families in Golbaf (total population: 16716) are participating in the pilot. Some of these families have begun raising the highly productive Afshar-Booroola sheep breed, to increase revenue and promote new management practices. Resalat Bank, through the establishment of a centralized Community Services Centre (the CSC), also supports an internet-based virtual space for all members, including vendors and service buyers. Any participating family that wants to set up a livestock production unit receives no-interest loans. Then, all actions related to buying, breeding, marketing, and value chain management are monitored under the CSC (described in detail in Fig. 7.3). Currently, the Veterinary Organization of Iran restricts many types of animal farming, however for this project, the province's authorities decided to make an exception for the city of Golbaf. Livestock farming

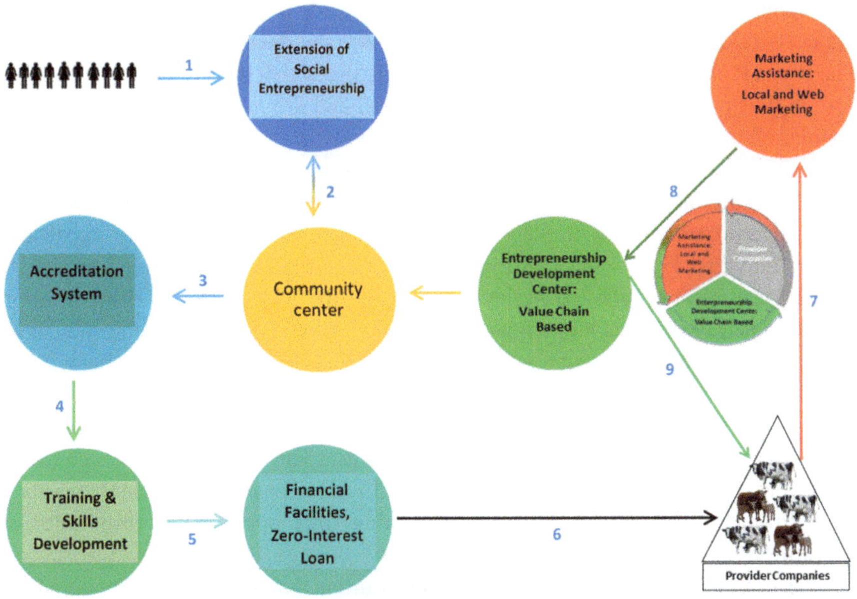

Fig. 7.3 Resalat Islamic Social Bank Support Structure for Golbaf Urban Sheep and Goat Raising Project. Notes: The structure of the project is as follows: 1—extension programs for introducing entrepreneurship to city communities. 2—eligible people are registered as members of Community Center which is the core of the organization, 3—aptitude assessment and accreditation system. In this step they would have a bank account for being a member of crowdfunding system 4—candidates enter to training and skills development program/courses and afterward are under supervision of technical mentor, 5—candidates introduce to bank for receiving zero-interest loan, 6—the loan finance is deposited to a provider company which is a part of integrated virtual market established by bank. In this step animals, feedstuffs, and other equipment are bought by candidates and start farming, 7, 8—farmers are helped to enter the value chain system, and 9—farmers sell their product in local and web based markets

in Golbaf is an example of urban farming, which currently supports food security of the city, as well as the creation of jobs and income, which contribute to poverty alleviation. A further benefit from the project is to stem out-migration from Golbaf to larger cities and to the capital.

On the Autobahn Between the Cities of Herne and Castrop-Rauxel (Germany): Urban Cattle to Manage Brownfields

In some urban settings, livestock production has been introduced in order to make use of the capacity of livestock to manage specific problems other than food provision. One such example is the recent introduction of the unique cattle breed Rotes

Höhenvieh (German Red) into a brownfield alongside a new highway construction project between two urban centers in western Germany (Herne and Castrop-Rauxel). The Rotes Höhenvieh are an older draught cattle breed that has not been competitive with higher producing breeds for milk and meat and are much reduced in terms of population numbers, almost to the point of extinction (Rischkowsky & Pilling, 2007). However, this breed is particularly good at landscape management and can graze less suitable sites more successfully than conventional breeds with low maintenance requirements (Schreiner, 2018). Developers for the highway expansion project between Herne and Castrop-Rauxel thus introduced this breed into a 21 hectare pasture alongside the highway as part of the environmental compensation for the highway project. The animals are managed by a local organic farmer in an extensive grazing system, into which 23 different species of plants have been introduced to convert what was once an industrial agricultural area into a more diverse ecological space. The herd is apparently growing and is well adapted to the brownfield space.[14] Animal products are sold by the farmer and there is no target end-date for the project, meaning that the herd will remain as long as it is possible to maintain the farm operation. As of the time of writing, this operation is profitable for the farmer.

This urban cattle pasture is part of a broader urban water management plan in the region to reclaim and re-naturalize the Emscher and Lippe waterways, which service one of the most densely populated areas in Germany (Perini & Sabbion, 2016). The cattle grazing in the brownfield help to manage the watershed by being part of a biodiverse "near-natural" landscape in the area versus the prior use of the waterway for a large open sewage system for the surrounding urban areas. The wastewater is now being managed underground,[15] leaving it possible to re-naturalize the above ground brownfield with cattle grazing and other systems.

German commitments to SDG 11 on sustainable cities, as well as their adopted policies to reduce conversion rates of natural areas to urban settlements and uses will necessitate increased innovative management of these brownfield areas in order to meet policy commitments (Masyk, 2021). Urban brownfields are also critical habitats for birds and other wildlife and can be a key pathway for the provision of ecosystem services into urban areas (Kamp et al., 2015; Mathey et al., 2015; Washbourne et al., 2020). An additional benefit of this project comes from preserving genetic diversity in cattle breeds, which is critical for maintaining healthy livestock populations in both urban and rural/conventional animal production systems (Curone et al., 2019; Tisdell, 2003). With an estimated 120,000-165,000 ha of brownfield space in Germany (Mathey et al., 2015), there is potential to scale up such initiatives for further environmental management as well as more developed urban livestock food production, even of such larger livestock breeds, like cattle.

[14] https://www.waz.de/staedte/herne-wanne-eickel/herne-rotes-hoehenvieh-an-der-a-42-hat-nachwuchs-bekommen-id232235223.html

[15] https://www.eglv.de/en/

Highlighted Features of Presented Urban Livestock Systems

The three case studies presented here showcase the high degree of variation in urban livestock production practices and motivations worldwide. Backyard poultry is growing in the city of Guelph due in large measure to urban consumer demand for these activities, a supportive city policy change allowing for more urban livestock keeping, as well as the setup of a host of supportive public, private, and academic institutions to guarantee animal health and welfare and to encourage best practices. In Golbaf, urban goat and sheep production is being taken up to respond to economic pressures from trade restrictions limiting food imports, as well as an overarching plan to use urban livestock production to stabilize smaller cities in terms of economic opportunity, limit out-migration, and boost food security. Urban cattle in western Germany in the Emscher watershed have been introduced mainly for non-food system purposes, making use of livestock for environmental management benefits.

However, despite the unique history and characteristics of these three urban livestock systems, some common elements can be observed. The first is that despite the wide variation in socioeconomic, agricultural, and policy environments across the three cities and countries we examined, urban dwellers in all three appear to have continued, consistent interest and demand for opportunities to engage in urban livestock production. Livestock products remain an important component of diets, and critical subsets of consumers in urban areas will pursue urban livestock production to supply some of this food demand. In the face of trade restrictions (Golbaf), this is an important way for residents to improve their food security and economic livelihoods in the face of sanctions-related limitations in accessing livestock products through conventional livestock value chains. Guelph residents' interest in backyard poultry may partly reflect global trends of urbanites who see such initiatives as environmentally sustainable, more humane, and who have an implicit skepticism of conventional livestock production and value chains (Blecha & Leitner, 2014). In other Canadian cities, urban poultry proponents have indicated that they wish to raise their own backyard hens in order to guarantee access to poultry and eggs due to COVID-19 pandemic-related supply disruptions and food insecurity,[16] although these expectations may sometimes diverge from realized poultry production costs in urban settings.[17] Urban livestock production systems and short value chains have proven to be resilient to a wide variety of shocks to conventional livestock systems. More structured investment in them can increase the food system resilience of urban

[16] Dempsey, K. (2021). "Backyard poultry proposal for Hanover" The Hanover Post, April 15, 2021 Retrieved October 4, 2021 from https://thepost.on.ca/news/local-news/backyard-poultry-proposal-for-hanover

[17] Ryan, D. (2020). "Backyard chickens are easy, but you may have them for their lifetime." July 20, 2020 in The Vancouver Sun. Retrieved on October 4, 2021 from https://vancouversun.com/news/backyard-chickens-are-easy-but-you-may-have-them-for-their-lifetime

areas (Marusak et al., 2021; Tittonell et al., 2021) and capitalize on this consistent urban consumer push for these systems.

Another broad feature of urban livestock systems is the need for many institutional supports to ensure animal and human health and welfare, access to veterinary services for urban animals and to provide information to newer urban farmers who do not have the training in animal production that conventional livestock producers do (Guelph and Golbaf). The cases studied also suggest the likely need for financing (Golbaf), as well as coordination with environmental and urban and development planning initiatives to identify ways to incorporate urban livestock production practices into overall planning objectives (Guelph, Golbaf, Herne). Current urban livestock activities are highly variable in the level of such supports, with resultant variation in urban livestock production outcomes. Best practices from conventional livestock production systems should be examined more rigorously to ensure that if urban livestock production is likely to become a more regular feature of urban food systems, animal welfare and environmental standards are maintained and tailored for unique urban environments.

Optimizing Urban Livestock Production

If urban livestock production is to be sustainable and to contribute to the food and environmental systems of the world's growing urban areas, it is clear that more efforts need to be made to translate traditional livestock production methods and best practices into more complex urban settings (O'Sullivan et al., 2019). Conventional livestock production systems are often based on spatially spread out systems with livestock production relegated to low population density rural areas, relying on improved transportation and market infrastructure to move livestock products to urban consumer centers. These systems are responsible for producing affordable livestock products to urban consumers worldwide and it is unlikely that urban livestock production systems will be able to compete with conventional systems that can capitalize on economies of scale and have a long history of research and investment in improved practices and efficiencies.

However, current research into conventional livestock production has also identified important negative environmental externalities associated with this approach, such as excess accumulation of animal waste, runoff into surrounding areas and waterways, and reduced nutrient cycling to intensive cropping systems due to transport costs (Russelle et al., 2007). Other research has highlighted potential negative animal welfare consequences to some intensive conventional livestock production practices that are increasingly questioned by urban consumers.[18] And finally,

[18] DAFA 2012. The Livestock Expert Forum. Science, economy, society—working jointly towards improvements in animal husbandry Strategy of the German Agricultural Research Alliance. ISBN 978-3-86,576-091-3. Retrieved October 4, 2021 from https://www.dafa.de/ag-research-in-germany/

dependence on long and growing conventional livestock value chains also has risks, as can be seen in the recent experience of the COVID-19 pandemic, which disrupted perishable livestock value chains worldwide, with resultant threats to farmers and urban consumers alike (Rivera-Ferre et al., 2021).

Fortunately, research is building into ways to minimize negative externalities and risks in conventional livestock production that may also have potential in urban areas. For example, integrated crop-livestock systems (ICLS) minimize environmental costs of livestock production by allowing animals to eat locally available food and fertilize crops in situ, rather than outsourcing both, often from synthetic sources requiring energy to manufacture and transport. Hou et al. (2008) highlight a cross-over of these principles to urban settings by examining successful integrated crop-livestock systems in urban areas in China, which follow ICLS best practices but use a diversity of animals suited to urban environments, including ponds for ducks and fish. The welfare implications of urban livestock production versus conventional livestock production have not been fully investigated to date, but guiding principles such as the Five Freedoms framework[19] for animal welfare can and should be applied to urban livestock production systems to guarantee that these systems abide by long-standing welfare standards and contribute positively to urban food systems and minimize harms to animals and humans. Some small-scale urban and peri-urban livestock production systems have proven to be resilient to livestock value chain shocks like the COVID-19 pandemic (Tittonell et al., 2021), but with high variability globally (Acosta et al., 2021). Thus, more research and efforts are needed to build up the resilience capacity of urban livestock production systems to make best use of these systems as complements to conventional systems, and also to capitalize on their best potential to contribute to urban food systems for the future.

References

Acosta, A., McCorriston, S., Nicolli, F., Venturelli, E., Aratchilage, U. G., ArceDiaz, E., et al. (2021). Immediate effects of COVID-19 on the global dairy sector. *Agricultural Systems, 103177*. https://doi.org/10.1016/j.agsy.2021.103177

Bijl, D. L., Bogaart, P. W., Dekker, S. C., Stehfest, E., de Vries, B. J., & van Vuuren, D. P. (2017). A physically-based model of long-term food demand. *Global Environmental Change, 45*, 47–62. https://doi.org/10.1016/j.gloenvcha.2017.04.003

Bijl, D. L., Bogaart, P. W., Dekker, S. C., & van Vuuren, D. P. (2018). Unpacking the nexus: Different spatial scales for water, food and energy. *Global Environmental Change, 48*, 22–31. https://doi.org/10.1016/j.gloenvcha.2017.11.005.Bruin2021

Blecha, J., & Leitner, H. (2014). Reimagining the food system, the economy, and urban life: New urban chicken-keepers in US cities. *Urban Geography, 35*(1), 86–108. https://doi.org/10.1080/02723638.2013.845999

[19] Farm Animal Welfare Committee (FAWC). (2016). Sustainable Agriculture and Farm Animals Welfare. Special Report by the Farm Animal Welfare Committee, Department for Environment, Food and Rural Affairs (Defra) UK, London. https://www.gov.uk/government/collections/fawc-advice-to-government

Brinkley, C., & Kingsley, J. S. (2018). Urban agriculture. In *Advances in Agricultural Animal Welfare* (pp. 241–257). Woodhead Publishing. https://doi.org/10.1016/B978-0-08-101215-4.00013-4
Capello, R. (2021). Classical contributions: Von Thünen and Weber. *Handbook of regional science*, 61–80. https://doi.org/10.1007/978-3-662-60723-7_94.
Curone, G., Filipe, J., Cremonesi, P., Piccioli-Cappelli, F., Trevisi, E., & Amadori, M. (2019). Relevance of the dairy cow biodiversity in the development of a profitable and environmentally sustainable livestock. *CAB Reviews, 14*(024), 1–11.
de Bruin, S., Dengerink, J., & van Vliet, J. (2021). Urbanisation as driver of food system transformation and opportunities for rural livelihoods. *Food Security, 1-18*. https://doi.org/10.1007/s12571-021-01182-8
Hou, F., Nan, Z., Xie, Y., Li, X., Lin, H., & Ren, J. (2008). Integrated crop-livestock production systems in China. *The Rangeland Journal, 30*, 221–231. https://doi.org/10.1071/RJ08018
Jamini, D., Amini, A., Gadermarzi, H., & Tavakoli, J. (2017). Challenges of food security in rural areas using grounded theory approach in ravansar county, western Iran. *European Journal of Geography, 8*(4), 26–40.
Kamp, J., Pelster, A., Gaedicke, L., Karthäuser, J., Dieker, P., & Mantel, K. (2015). High nest survival and productivity of northern lapwings Vanellus vanellus breeding on urban brownfield sites. *Journal of Ornithology, 156*(1), 179–190. https://doi.org/10.1007/s10336-014-1114-0
Kheraj, S. (2015). Urban environments and the animal nuisance: Domestic livestock regulation in nineteenth-century Canadian cities. *Urban History Review/Revue d'histoire urbaine, 44*(1-2), 37–55. https://doi.org/10.7202/1037235ar
Marusak, A., Sadeghiamirshahidi, N., Krejci, C. C., Mittal, A., Beckwith, S., Cantu, J., et al. (2021). Resilient regional food supply chains and rethinking the way forward: Key takeaways from the COVID-19 pandemic. *Agricultural Systems, 190*, 103101. https://doi.org/10.1016/j.agsy.2021.103101
Masyk, T. A. (2021). The role of urban agriculture for sustainable and resilient cities (Doctoral dissertation, TU Dortmund University, Germany). https://eldorado.tu-dortmund.de/bitstream/2003/40281/1/20210615-Dissertation-Masyk-V8_PRINT.pdf
Mathey, J., Rößler, S., Banse, J., Lehmann, I., & Bräuer, A. (2015). Brownfields as an element of green infrastructure for implementing ecosystem services into urban areas. *Journal of Urban Planning and Development, 141*(3), A4015001. https://doi.org/10.1061/(ASCE)UP.1943-5444.0000275
Michaels, G., Rauch, F., & Redding, S. J. (2012). Urbanization and structural transformation. *The Quarterly Journal of Economics, 127*(2), 535–586. https://doi.org/10.1093/qje/qjs003
Mirjalili, S. H., Baan, H., Abouzar, S., & Botshekan, M. H. (2020). Islamic social banking platform (case of Resalat Islamic Bank). 10.29252/jme.15.1.35. *Journal of Money and Economy, 15*(1), 35–54.
Mohammadi, J., & Soleimani, S. M. (2014). The effect of urban agriculture in urban sustainable development and its techniques: A case study in Iran. *International Journal of Agriculture and Forestry, 4*(4), 275–285.
O'Sullivan, C. A., Bonnett, G. D., McIntyre, C. L., Hochman, Z., & Wasson, A. P. (2019). Strategies to improve the productivity, product diversity and profitability of urban agriculture. *Agricultural Systems, 174*, 133–144. https://doi.org/10.1016/j.agsy.2019.05.007
Perini, K., & Sabbion, P. (2016). *Urban sustainability and river restoration: Green and blue infrastructure*. John Wiley & Sons. https://doi.org/10.1002/9781119245025
Pilehvar, A. (2021). Spatial-geographical analysis of urbanization in Iran. *Humanities and Social Sciences Communications, 8*(1), 1–12.
Rischkowsky, B., & Pilling, D. (2007). *The state of the world's animal genetic resources for food and agriculture*. Food & Agriculture Org.
Ritchie, H., & Roser, M. (2018). Urbanization. *Our world in data*. Retrieved from: https://ourworldindata.org/urbanization [Online Resource].
Rivera-Ferre, M. G., López-i-Gelats, F., Ravera, F., Oteros-Rozas, E., di Masso, M., Binimelis, R., & El Bilali, H. (2021). The two-way relationship between food systems and the COVID19

pandemic: Causes and consequences. *Agricultural Systems, 191*, 103134. https://doi.org/10.1016/j.agsy.2021.103134

Russelle, M. P., Entz, M. H., & Franzluebbers, A. J. (2007). Reconsidering integrated crop–livestock systems in North America. *Agronomy Journal, 99*(2), 325–334.

Schreiber, K., Hickey, G. M., Metson, G. S., Robinson, B. E., & MacDonald, G. K. (2021). Quantifying the foodshed: A systematic review of urban food flow and local food self-sufficiency research. *Environmental Research Letters, 16*(2), 023003.

Schreiner, J. A. (2018). Saving the breeds: German farmers' preferences for endangered dairy breed conservation programs. *Proceedings in System Dynamics and Innovation in Food Networks, International Journal on Food System Dynamics, 2018*, 182–193. https://doi.org/10.18461/pfsd.2018.1813

Seto, K. C., & Ramankutty, N. (2016). Hidden linkages between urbanization and food systems. *Science, 352*(6288), 943–945.

Tisdell, C. (2003). Socioeconomic causes of loss of animal genetic diversity: Analysis and assessment. *Ecological Economics, 45*(3), 365–376. https://doi.org/10.1016/S0921-8009(03)00091-0

Tittonell, P., Fernandez, M., El Mujtar, V. E., Preiss, P. V., Sarapura, S., Laborda, L., Mendonça, M. A., Alvarez, V. E., Fernandes, G. B., Petersen, P., & Cardoso, I. M. (2021). Emerging responses to the COVID-19 crisis from family farming and the agroecology movement in Latin America—A rediscovery of food, farmers and collective action. *Agricultural Systems, 190*, 103098. https://doi.org/10.1016/j.agsy.2021.103098

Varga, C., Guerin, M. T., Brash, M. L., Slavic, D., Boerlin, P., & Susta, L. (2019). Antimicrobial resistance in fecal Escherichia coli and salmonella enterica isolates: A two-year prospective study of small poultry flocks in Ontario, Canada. *BMC Veterinary Research, 15*(1), 1–10. https://doi.org/10.1186/s12917-019-2187-z

Washbourne, C. L., Goddard, M. A., Le Provost, G., Manning, D. A., & Manning, P. (2020). Trade-offs and synergies in the ecosystem service demand of urban brownfield stakeholders. *Ecosystem Services, 42*, 101074. https://doi.org/10.1016/j.ecoser.2020.101074

Chapter 8
Organizing Urban Farmers in a Post-COVID-19 Economic Recovery

M. Donald Houessou, Ben G. J. S. Sonneveld, Augustin K. N. Aoudji, and Amani Alfarra

Abstract The COVID-19 pandemic hit the food supply chains hard, especially in urban areas where inhabitants mostly depend on market-based food systems. Despite the intended exemptions made for food transport reality shows different in that the imposed mobility restrictions by COVID-19 prevention measures interrupted both agricultural and food supply chains. The symptoms of the impact were most visible for individual and dominant organizational forms of urban farms who could not levy negotiation power to buy agricultural inputs and had to forego sales to their regular customers. Inversely, joint ventures of urban farmers that buy inputs and sell produce collectively were generally a success because their sale force fosters the requests to the official authorities for food transport and distribution. Therefore, COVID-19 mobility restrictions improved collaborations among urban farmers who experienced the clear benefits of cooperative structures on their livelihoods. Such positive effects should inform agricultural advices curriculum on farmers' organization to improve the ways urban farmers cooperate in a post-COVID-19 recovery.

Keywords Agricultural economics · Supply chains · Urban agriculture

M. D. Houessou
The African Center for Equitable Development (ACED), Quartier Tokan, Abomey Calavi, Benin

B. G. J. S. Sonneveld (✉)
Amsterdam Centre for World Food Studies, Vrije Universiteit Amsterdam, De Boelelaan, Amsterdam, Netherlands
e-mail: b.g.j.s.sonneveld@vu.nl

A. K. N. Aoudji
Faculty of Agricultural Sciences of the University of Abomey-Calavi, Abomey-Calavi, Republic of Benin

A. Alfarra
Food and Agriculture Organization of the United Nations, Land and Water (NSL) Division, Viale delle Terme di Caracalla, Rome, Italy

Walter G. Booth School of Engineering Practice and Technology, Hamilton, ON, Canada
e-mail: Amani.Alfarra@fao.org

A. A. Kubursi et al. (eds.), *Urban Food Security in a Crisis Prone World*,
https://doi.org/10.1007/978-3-031-89440-4_8

Why Organizing Urban Farmers Matters?

The rapid expansion of the coronavirus disease 2019 (COVID-19) pandemic has aggravated risks of severe food insecurity from 135 million in January 2020 to 265 million by the end of 2020 (Dongyu, 2020). This food insecurity problem will become worse if new waves of COVID-19 contamination continue to appear. Indeed, the COVID-19 pandemic has triggered some strict mobility restrictions measures that disrupted the food supply chains worldwide limiting production and food availability, especially in countries that depend on food import (Almohamad et al., 2020). In urban areas, the COVID-19 pandemic revealed the extreme fragility of cities to unexpected complex global risks as densely populated cities are sites of consumption, almost completely dependent on outside resources for food flows (Pulighe & Lupia, 2020).

The fragility of cities to food insecurity during lockdowns has led to an increasing awareness of the importance of producing essential goods such as food in urban areas (Pulighe & Lupia, 2020). This call brings to the contribution of urban agriculture (UA) in feeding cities with fresh and healthy foods. Overall, the global popular and scientific discourse—from developed and developing countries—is positive on the necessity to promote UA to transform and diversify food systems and increase the resilience of cities against food insecurity problems (Pulighe & Lupia, 2020; Mead et al., 2021; Houessou et al., 2021; Cockburn, 2020). However, although not obvious, the way to sustainably match short food supply lines with local demand is a great challenge to be addressed but the literature largely ignores this subject (Pulighe & Lupia, 2020; Houessou et al., 2021).

This chapter aims to address the knowledge gap and learnt from the case of the Republic of Benin where imposed lockdown in cities impede urban farmers to distribute their production. Benin is a West-African country with 12 million inhabitants and a gross domestic product per capita of USD 1291 in 2020 (World Bank, 2021). The country witnesses a stunning growth of its urban population, passing from 43% in 2010 to 48% in 2020 (World Bank, 2021), and the emergence of allotment gardens as the most practiced form of UA (Houessou et al., 2019a). The country is also facing the challenges brought about by the COVID-19 pandemic and related measures on food security, especially in urban areas. Against the difficulties of buying inputs and selling produce, some urban farmers improved their level of cooperation and accessed market opportunities that would have been impossible with the individual. Lessons from this case would be useful to inform a post-COVID-19 economic recovery on how to organize urban farms to success.

The chapter is organized as follows. First the low level of organization among urban farmers is discussed, learning from the case of Benin. Next, the case of Benin is further elaborated and shows how the COVID-19 pandemic triggered a better cooperation among some urban farmers. Finally, the findings are summarized and some recommendations are formulated that foster collective actions among urban farms in the future post-COVID-19 world.

The Low Collective Actions among Urban Farmers

In general, collective actions, regardless of the agricultural subsector, among farmers are generally low in developing countries (Devaux et al., 2009; Houessou et al., 2019b; Arouna, 2018; Abate, 2018). Collective action refers to voluntary action taken by a group to pursue common interests or achieve common objectives. In collective action, members may act on their own, but more commonly they act through a group or an organization; they may act independently or with the encouragement or support of external agents from governmental bodies, nongovernmental organizations (NGOs), or development projects (Meinzen-Dick & Di Gregorio, 2004). Exploring this low organization rate among urban farmers, this section drew from a previous investigation conducted in 2019 among 261 urban gardeners in two cities—Cotonou and Porto-Novo—in Benin. The investigation explored various aspects that might explain successes or failures in the organization and functioning of cooperative structures that prevail in urban allotment gardens (Houessou et al., 2019b).

The investigation showed that urban gardeners usually express the need to collaborate within their groups. Therefore, they are mostly organized in cooperatives and regularly subscribe (weekly, monthly, and yearly) to a membership fee. The membership fee is intended to purchase fertilizer and seeds, but in practice gardeners individually purchase their inputs. The cooperatives in most of the cases have a management team, elected in 50% of the cases. Gardeners also recognize the authority of the management team, share the responsibility for common tasks (site surveillance, participation in meetings, and offering help to other members) and have a constitution. The cooperatives are also registered with the ministry of agriculture and have bank accounts that collect membership fees. Besides purchases, the fees are also intended to provide guarantees for taking credit but, in general, this is not effective due to a lack of trust between members. The cooperatives also collaborate with their communal and national representations that defend their interests during policy dialogues. By comparing this situation of urban gardeners in Benin with the seven principles of successful cooperatives put in place by the International Cooperatives Alliance (ICA) (Table 8.1), it appears that cooperatives are well structured in principle but still struggle to set clearly defined rules that drive collective actions: agreement on rules of ownership, member control, economic participation, autonomy, and concern for community.

Table 8.1 Comparison between ICA principles for cooperatives and urban gardeners' organizations

ICA cooperative principles	29 gardeners' organizations
(1) Voluntary and open membership	Present
(2) Democratic member control	Present in some
(3) Member economic participation	Present in some
(4) Autonomy and independence	Present in some
(5) Education, training, and information	Present in some
(6) Cooperation among cooperatives	Present
(7) Concern for community	Not present

Source: Adapted from Kurimoto, Draperi (Kurimoto et al., 2015)

Concretely, the investigation found that very few gardeners form joint partnerships to purchase inputs and sell produce that would probably increase their transaction costs. Although the mistrust among members was indicated as one of the reasons, urban gardeners who are dissatisfied with individual marketing experiences are willing to enter joint ventures if, in general, the issues of mistrust regarding members and the management team are tackled. For instance, there were some few examples of joint ventures where gardeners were successful as a group to obtain loans from financial institutions to boost their activities. These examples corroborate the argument that collective actions may actually assure a common interest and provide assurance to a wide group of involved people. The examples also open the debate for financial incentives that may encourage urban gardeners to make important business decisions collectively. In that perspective, support from national economic policies may be needed to facilitate access to loans for farmers which, in return, may improve their collective actions and business.

The Positive Side of the COVID-19 Pandemic in Collective Actions

After the country identified its first case, the Beninese government took a series of barrier measures of progressive intensity, one of which was the establishment of a sanitary cordon or "cordon sanitaire" to isolate the high-risk zones from other regions of the country. Specifically, a sanitary cordon is the restriction of movement of people into or out of a defined geographic area, such as a community, region, or country (Rothstein, 2015). It is generally created around an area experiencing an epidemic or an outbreak of infectious disease or along a border between nations. Once established, people from the affected urban areas were no longer allowed to leave or enter the sanitary cordon but, only the transport and distribution of agricultural and food products were allowed if official authorizations were obtained. Despite the food exemptions, urban gardeners experienced some difficulties in accessing inputs and selling their produce and only a tiny groups of gardeners could organize to access the market opportunities. This section learns from a recent policy dialogue held in February 2021 in Benin with local and national policymakers, civil society organizations, private sector, and gardeners association. The policy dialogue aimed to discuss the effects of the COVID-19 pandemic on food systems (ACED, 2021).

The effects of the COVID-19 pandemic on the food systems were visible in the country. First, there was a widespread hike in food prices despite the food exemptions made during the imposed movement restrictions. Additionally, food distribution to consumers and inputs supply to gardeners were reduced, which exerted an extra demand pressure on food, resulting in price increases over again. This situation was even more pronounced in areas outside the cordon sanitaire, since the cordon was concentrated on the major urban centers and thus included most suppliers of inputs, seeds, and phytosanitary products. As a result, many supply contracts for gardeners were terminated as they could not organize to supply food to consumers. Indeed, although food transport was exempted from the mobility restrictions, it was an obligation for gardeners or traders to request an official authorization before food movements. Such an obligation could only be met by some groups of gardeners who were able to aggregate their produce to supply consumers or their needs to buy inputs. The many gardeners who were operating individually could not obtain such authorizations to make transactions and continue their agricultural business; hence, a decline in their food security and living conditions. These empirical observations seem to demonstrate that movement restrictions triggered among urban gardeners the necessity to improve their collaborations and make collective actions—as initially desired when creating the groups—to access market opportunities and achieve common goals.

Way Forward to Organize Urban Farmers in a Post-COVID-19 Recovery?

This chapter aimed to respond to the way to sustainably match short food supply lines with local demand and discussed the level of cooperation among urban gardeners as a solution. Using the case of urban gardeners in Benin, the chapter showed that the level of organization within their cooperatives is low and deserves some improvements. Against the mobility restrictions imposed by the COVID-19 pandemic and the incapacity to buy inputs and sell produce individually, some groups of gardeners were able to improve their collaborations and requested official authorizations to access market opportunities. These successful examples that used the cooperatives power as a negotiation force to make collective actions clearly demonstrated the benefits associated with the development of genuine cooperatives among urban farmers on livelihoods and the society alike. For instance, previous successful examples of collective actions on rural farmers' income can be found in Benin within cashew nuts farmers (Mensah et al., 2012) and paddy rice producers (Arouna, 2018); in Namibia (Thomas & Hangula, 2011), South Africa (Ortmann & King, 2007), and Kenya within dairy farmers (Staal et al., 1997); Ethiopia (Abate, 2018) and the Andean region of South America within potato farmers (Devaux et al., 2009). Although, these examples are mostly for commercial and staple products, they provide evidence that farmers' commitment in collective actions can improve access to market opportunities and lives.

By looking at the ICA's principles, it becomes clear that more efforts should be put on rules that define common resources' sharing, control of management board by members, and economic participation of members and revenues retribution to members. These efforts should start by the urban farmers themselves as a commitment to succeed collectively. Then, specialized nongovernmental organizations can enter the room and support the process by building or strengthening the capacity of groups of farmers on group management. Two key issues that would need special attention during these capacity building programs are financial management and conflict management, as also recommended by some previous studies that analyzed important factors affecting proper functioning of smallholder cooperatives in Benin and South Africa (Arouna, 2018; Thaba et al., 2016). Moreover, the support of agricultural and economic policies is necessary by establishing a guarantee fund, which may encourage financial institutions to take more risks in agriculture financing, thereby enticing urban farmers to improve their cooperation. This policy support is also identified by previous studies as important to create an enabling policy environment for urban farmers (Hellin et al., 2009; Gning & Larue, 2014). Finally, lessons learnt from this chapter would be useful to inform a strategy on cooperatives development in a post-COVID-19 economic recovery.

References

Abate, G. T. (2018). Drivers of agricultural cooperative formation and farmers' membership and patronage decisions in Ethiopia. *Journal of Co-operative Organization and Management.*

ACED (2021). *Dialogue politique sur la promotion de l'agroécologie et du consommer local au Bénin.*, A. IPADEV, Editor, ACED & IPADEV: https://www.aced-benin.org/publication/rapport-du-dialogue-politique-sur-la-promotion-de-lagroecologie-et-du-consommer-local-au-benin/. p. 16.

Almohamad, M., Mofleh, D., & Sharma, S. (2020). The impact of the COVID-19 pandemic on food insecurity. *Journal of Agriculture, Food Systems, and Community Development, 10*(1), 1–4-1–4.

Arouna, A. (2018). Assessing the impact of collective Marketing of Paddy Rice in innovation platforms by smallholder producers in Benin. In *Rice Crop-Current Developments*. IntechOpen.

Cockburn, H. (2020). *As gardeners turn to growing own food, research reveals dramatic decline in urban allotments over last 50 years*. Retrieved March 27, 2020, from https://www.independent.co.uk/climate-change/news/grow-food-coronavirus-urban-allotments-fruit-vegetables-a9431051.html.

Devaux, A., et al. (2009). Collective action for market chain innovation in the Andes. *Food Policy, 34*(1), 31–38.

Dongyu, Q. (2020). *Senior officials sound alarm over food insecurity, warning of potentially 'biblical' famine, in briefings to security council title. United Nations Security Council.* , S.C.A. Division, Editor. https://www.un.org/press/en/2020/sc14164.doc.htm.

Gning, T., & Larue, F. (2014). *Le nouveau modèle coopératif dans l'espace OHADA: un outil pour la professionnalisation des organisations paysannes?* Fondation pour l'agriculture et la ruralité dans le monde (FARM).

Hellin, J., Lundy, M., & Meijer, M. (2009). Farmer organization, collective action and market access in Meso-America. *Food Policy, 34*(1), 16–22.

Houessou, D., et al. (2019a). Atif Kubursi et al. (Eds): Urban Food Security in a Crisis Prone World *Urban agriculture in Benin: How can policy support gardeners?* Report on a survey conducted-among experienced urban gardeners in Cotonou and Porto-Novo, Benin. Centre d'Actions pour l'Environnement et le Développement Durable (ACED), Benin, Amsterdam Centre for World Food Studies/Athena Institute, Vrije Universiteit, The Netherlands, Faculty of Agricultural Sciences/University of Abomey-Calavi, Benin. https://doi.org/10.13140/RG.2.2.11319.57766

Houessou, D. M., et al. (2019b). How to transition from cooperations to cooperatives: A case study of the factors impacting the organization of urban gardeners in Benin. *Sustainability, 11*(17), 4528.

Houessou, M. D., Cassee, A., & Sonneveld, B. G. (2021). The effects of the COVID-19 pandemic on food security in rural and urban settlements in Benin: Do allotment gardens soften the blow? *Sustainability, 13*(13), 7313.

Kurimoto, A., et al. (2015). Guidance notes to the co-operative principles. *Consultado em, 10*(01), 2016.

Mead, B. R., et al. (2021). Urban agriculture in times of crisis: The role of home food growing in perceived food insecurity and well-being during the early COVID-19 lockdown. *Emerald Open Research, 3*(7), 7.

Meinzen-Dick, Ruth Suseela, ed.; Di Gregorio, Monica, ed. 2004. *Collective action and property rights for sustainable development.* 2020 Vision Focus 11. Washington, DC: International Food Policy Research Institute (IFPRI). https://hdl.handle.net/10568/157457

Mensah, E. R., et al. (2012). Determinants of commitment to agricultural cooperatives: Cashew nuts farmers in Benin. In *International association of agricultural economists triennial conference. Foz Do Iguaçu, Brazil.*

Ortmann, G. F., & King, R. P. (2007). Agricultural cooperatives I: History, theory and problems. *Agrekon, 46*(1), 18–46.

Pulighe, G., & Lupia, F. (2020). Food first: COVID-19 outbreak and cities lockdown a booster for a wider vision on urban agriculture. *Sustainability, 12*(12), 5012.

Rothstein, M. A. (2015). From SARS to Ebola: Legal and ethical considerations for modern quarantine. *Indian Health Law Review, 12*, 227.

Staal, S., Delgado, C., & Nicholson, C. (1997). Smallholder dairying under transactions costs in East Africa. *World Development, 25*(5), 779–794.

Thaba, K., Anim, F., & Tshikororo, M. (2016). Analysis of factors affecting proper functioning of smallholder agricultural cooperatives in the Limpopo Province of South Africa. *Journal of Human Ecology, 54*(3), 150–157.

Thomas, B., & Hangula, M. M. (2011). Reviewing theory, practices and dynamics of agricultural cooperatives: Understanding cooperatives development in Namibia. *Journal of Development and Agricultural Economics, 3*(16), 695–702.

World Bank (2021). *World Bank national accounts data: Benin.* https://data.worldbank.org/indicator/NY.GDP.PCAP.CD?locations=BJ.

Chapter 9
Selecting Allotment Gardens in Urban and Peri-Urban Areas

Ben G. J. S. Sonneveld, M. Donald Houessou, Augustin K. N. Aoudji, and Amani Alfarra

Abstract The downside of the current rapid urbanization is well known; cities grow faster than their minimum service levels can cope with. This is particularly manifested in urban food insecurity, predominantly in the lower wealth segments of the Global South countries, which demands far reaching policy interventions to empower the urban poor and to allow them to tackle these problems heads-on. The impoverished urbanized population, the world over, faces now even greater challenges under the COVID-19 mobility restrictions that disrupt food supply chains while those who lose their jobs in the informally organized sector do not qualify for income compensation. Urban agriculture seems a viable answer; it can be practiced at safe and infectious preventive distances, produces healthy and nutritious food, earns some extra income from surplus production, and can be located away from infection prone mass concentrations. Political support to implement allotment gardens is largely absent and continues to be sustained by information gaps on the allocation and suitability of potential sites.

We address this knowledge gap by presenting the fundamental building blocks for a site allocation tool that optimizes the spatial planning of allotment gardens in urban and peri-urban areas while accounting for biophysical (soil, water) and socioeconomic (distance to markets and safe entrance for women) conditions as well as COVID-19

B. G. J. S. Sonneveld (✉)
Amsterdam Centre for World Food Studies (ACWFS), Vrije Universiteit Amsterdam, De Boelelaan, Amsterdam, Netherlands
e-mail: b.g.j.s.sonneveld@vu.nl

M. D. Houessou
The African Center for Equitable Development (ACED), Quartier Tokan, Abomey Calavi, Benin

A. K. N. Aoudji
Faculty of Agricultural Sciences of the University of Abomey-Calavi, Abomey-Calavi, Republic of Benin

A. Alfarra
United Nations Food and Agriculture Organization, Rome, Italy

A. A. Kubursi et al. (eds.), *Urban Food Security in a Crisis Prone World*,
https://doi.org/10.1007/978-3-031-89440-4_9

safety measures and will provide better life conditions. The tool underlying the model uses a wide array of fully georeferenced database of explanatory variables while a selected of subset expert judgments assess site suitability in qualitative ordered classes. Model outcomes are tested for consistency, reproducibility, and stability (robustness), while a correlation protocol with quantitative data makes qualitative judgments interpretable. The model is converted into a tool for an impact interventionist analysis and aims to foster sufficient production and consumption of food locally in urban settings.

Keywords Allotment gardens · Food security · Peri-urban · Urban agriculture

Introduction

Propelled by a high rate of population growth and a fast-reducing number of job opportunities, the rapid growth of ever-growing mega poles in the Global South should not come as a surprise. Figure 9.1 shows the clear shift from a rural to urban society in Asia and Africa.

In South America, 15% of the population is now active in rural areas, which will decline to 10% in 2050 (FAO, 2021). Yet, the rapid urban expansion also comes at a price as the number of new residents outstrips the city's capacity to deliver basic services. Many urban dwellers in the Global South end up in slums where they lack access to clean water and sanitation, reside in houses with insufficient living areas, and suffer from informal and unguarded tenure security (World Bank, 2021). Figure 9.2 shows that a stunning 54% of the urban population in Sub-Saharan Africa lives in slums while in South Asia the urban population ranks a close second with

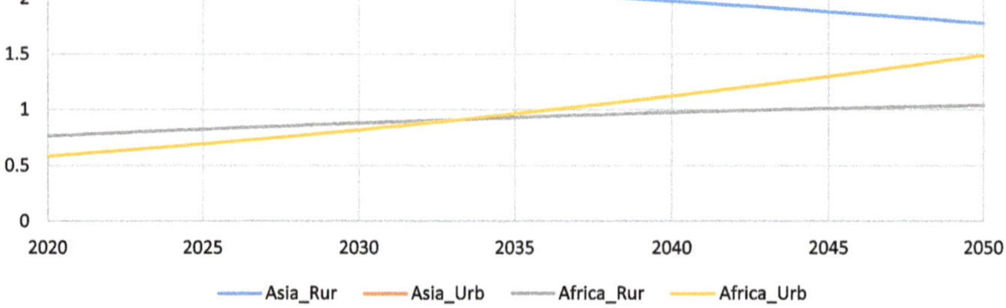

Fig. 9.1 Future urban and rural population (in billions) in Africa and Asia over the period 2020–2050

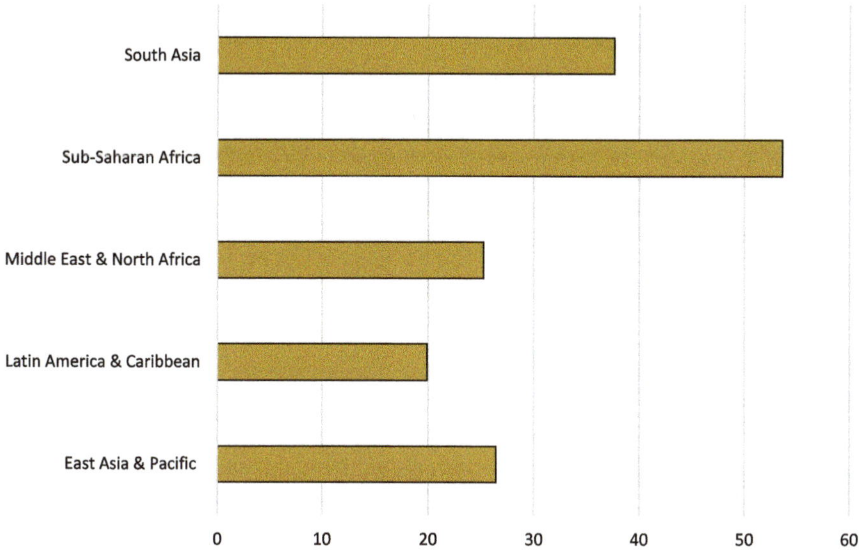

Fig. 9.2 Share of urban population that lives in slum areas

38%. The uncontrolled growth of cities also affects the biotope in which the population has to survive. For example, unplanned sewage systems that flow through the slums affect water quality causing water-borne diseases while air pollution impacts human health. Additionally, the Intergovernmental Panel on Climate Change demands integral solutions to cut the fast rising carbon emissions from urban and peri-urban areas. Health hazards caused by inadequate basic services raise the risk of communicable diseases, which became particularly evident during the COVID-19 pandemic (Rocklöv & Sjödin, 2020). High-density housing with a large number of inhabitants per dwelling combined with feeble sanitation services impedes a successful implementation of containment strategies in slums (Friesen & Pelz, 2020). Yet, the impact of measures that address the COVID-19 pandemic reached further than the infection hazard. Due to lockdown measures many inhabitants of slum areas lost their jobs and income and were forced to lower their food consumption. Access to promised income compensation schemes often failed (Houessou et al., 2021), though agility and knowledge about slums enabled some NGO's to give a rapid response to the crisis (Pongutta et al., 2021).

In a rapid food security appraisal, Houessou et al. (2021) conducted among 240 rural and urban dwellers in southern Benin a study that evaluated the mobility restrictions and food security of the COVID-19 imposed "cordon sanitaire." The raging pandemic affected the four food security pillars (availability, accessibility, utilization, and stability) in both rural and urban areas and within and outside the cordon sanitaire. Respondents inside the cordon sanitaire, specifically rural producers and urban inhabitants without access to allotment gardens were among the most affected (Fig. 9.3). Reasons for the decline most frequently cited were: increased food prices, disruptions in food logistics, and loss of work and income due to

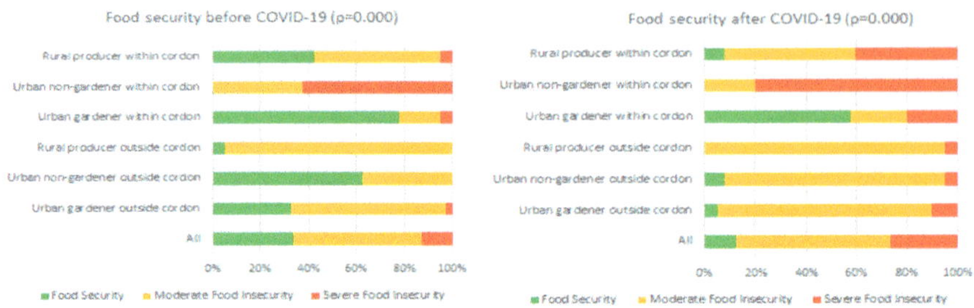

Fig. 9.3 Food security situation of the respondents

movement restrictions. Access to allotment gardens mitigated the food crisis during the pandemic.

The above example confirms the imminent role of urban gardens and urban agriculture in preserving food security especially in low- and middle-income countries like Gabon (Agnes et al., 2016; Bayendi-Loudit et al., 2017), Democratic Republic of Congo (Musibono et al., 2011) and Malaysia (Rezai et al., 2016). Moreover, UA contributes to a circular economy at the city level creating sustainable and resilient urban communities (Ferreira et al., 2018). Furthermore, UA mitigates negative climate change impacts by creating buffer zones where the excess water can infiltrate and be stored for periods of drought. The green spaces created by UA also form a powerful response to "heat islands" within the city areas. For these reasons, there are increasing calls to incorporate UA into urban planning to create city-wide sustainability (Langemeyer et al., 2021) that places and maintains a natural resources base to improve the overall quality of life and reduce negative climate impacts. As Lee (2021) stated, "Urban gardens can serve as a connection to the natural world, a sustainable food source, and may even bring communities closer together."

Concerning Sustainable Development Goals, urban agriculture contributes to SDG 11 -"Make cities and human settlements inclusive, safe, resilient and sustainable," but also to SDG 2 "End hunger, achieve food security and improved nutrition and promote sustainable agriculture," SDG 6 "Ensure availability and sustainable management of water and sanitation," SDG 8 "Sustainable economic growth, full employment and decent work for all," SDG 12 "Ensure sustainable consumption and production patterns," SDG 13 "Take urgent action to combat climate change and its impacts," and; SDG 15 "Protect, restore and promote sustainable use of terrestrial ecosystems."

However, despite the enthusiasm and willingness to start and participate in urban garden projects urban planners are often ill-equipped to inform the city administrations on possibilities for a multifunctional urban garden project that respects the prevailing natural resource base, matches specific needs and preferences of local residents, and is planned for the use by several generations.

This chapter addresses these concerns by introducing the principal steps for the creation of a site selection tool that accounts for local situations and the specific desired conditions imposed by the practitioners. Hence, allocation of urban gardens is not only dependent on the natural resource base like fertile and tillable soil and good quality water, it also relies on socioeconomic considerations as distance to markets and proximity of homestead and health aspects like the omnipresence of communicable diseases like the COVID-19. Hence, the protocol that describes the allocation tool provides a platform for the formal integration of these important factors to influence the decision-making processes.

The Protocol

The objective of the planning tool is to inform policy-makers about the planning of urban allotment gardens in cities and peri-urban areas. The design and development of the site allocation tool consist of seven steps:

- Compilation of independent variables.
- Compilation of dependent variable: expert judgments.
- Preparation for estimation: conversion of maps into vector.
- Estimation protocol.
- Decision support tool: Assess suitability on unvisited sites for various options.
- Visualize results: Conversion from vector into map.
- Policy interpretation of outcomes.

The Seven Steps

In this section we introduce the seven steps that will lead to a site allocation tool.

Step 1. Compilation of Independent Variables

The compilation of variables that can be used to explain the suitability of the sites for urban gardens creates a multidisciplinary data base that reflects the aspirations of decision-makers to account for the specific selected factors.

Georeference. The information of independent variables is based on grid maps with equal georeference. Explicitly, maps with similar: projection, grid size, corner pixel, number of rows, number of columns.

Raster maps (Fig. 9.4) with a different projection can be resampled[1] to a common georeference. Grid cell resolutions higher than or equal to 30 × 30 m are recommended.

[1] Using resampling techniques like: nearest neighbor, bilinear, and bicubic.

```
ncols 2200
nrows 1700
xllcenter 2
yllcenter 6
cellsize 0.000278
nodata_value -32767
125 123 123 124 124 126 127 128 127 127 126 126 127 127 126 126 126 125 125 126
125 123 122 125 126 127 128 128 128 127 126 125 126 128 128 126 125 125 125 125
125 124 124 125 127 127 127 128 128 127 126 125 126 128 128 127 126 126 125 126
125 124 124 124 124 126 126 127 128 127 126 126 127 127 128 127 126 125 125 125
124 124 124 122 123 125 126 126 128 129 127 127 127 126 126 126 126 125 125 125
125 124 123 121 122 125 125 126 128 128 127 127 127 125 125 125 126 126 126 126
125 124 124 123 122 124 125 127 128 128 127 127 127 126 125 125 125 126 127 127
124 124 124 122 123 124 125 127 128 128 128 127 126 126 126 125 125 126 126 126
124 124 123 122 123 124 125 128 128 128 128 127 126 126 126 126 126 126 126 126
124 123 121 121 123 124 126 128 129 128 127 126 127 126 126 127 128 128 126 126
124 123 121 121 123 125 126 128 129 127 125 125 125 125 126 127 129 128 126 125
124 123 122 121 123 124 125 127 126 125 124 125 125 125 126 126 128 128 127 125
123 122 121 120 122 123 123 124 124 123 124 125 125 126 126 126 127 127 126 125
121 121 120 120 121 121 122 124 125 124 124 124 125 125 125 126 126 126 126 125
121 120 119 120 121 121 123 126 126 125 124 124 125 126 125 126 126 125 125 125
120 120 119 120 121 122 124 126 126 125 124 123 125 125 125 125 126 126 126 127
120 121 120 120 121 122 124 126 126 125 124 123 124 125 125 124 125 125 126 126
120 121 121 121 121 122 124 126 126 125 124 124 125 126 126 125 125 125 125 125
119 120 121 121 121 122 124 126 125 125 124 125 126 126 127 126 126 125 125 125
120 119 119 119 120 120 124 125 126 126 125 125 125 125 126 126 125 125 125 126
121 119 118 118 119 119 122 123 124 126 126 125 124 124 124 124 124 124 125 126
122 119 118 118 119 119 121 121 123 125 126 125 122 122 122 123 124 125 126 127
121 119 118 118 120 120 120 120 122 125 126 125 122 121 122 124 126 126 127 128
120 118 117 118 120 120 120 121 122 124 124 125 123 122 123 125 127 126 126 127
119 118 117 118 119 119 120 120 121 122 122 123 123 125 126 127 128 127 127 127
118 118 118 119 119 118 120 121 119 119 120 122 123 125 128 128 128 128 128 129
119 118 119 119 119 119 121 121 119 118 119 121 122 124 125 127 127 127 128 129
```

Fig. 9.4 A raster map with specific georeference (ASCI) read in a Text Editor

Crossing. Raster maps of the variables can be crossed and result in a multivariate representation for each grid cell. Moreover, raster maps are easy to read and can be converted into formats for statistical analysis (see also step 3).

Attributes. Commonly the map attributes representing the explanatory variables are subdivided into the following main categories with the map attributes.

Biophysical data	Agriculture	Socioeconomic data	Marketing	Health
Climate	Crop water requirements	Population density	Distance to markets	Distance to hospitals
Water resources	Crops nutrient requirements	Urban/Rural areas	Prices	COVID-19 infection concentrations
Soil	Irrigation	Cadastre	Cooperatives	
Land use	Landscaping/ Tillage	Distance to homestead		
Geology	Land use	Income distribution		
		Forensic information: e.g., Robberies		

Step 2. Compilation of Dependent Variable: Expert Judgments

Expert judgments on site suitability are a valuable source of information in areas where data paucity prevails. The difficulties of applying empirical models in earth sciences are well acknowledged (Verheye & Dent, 1997) and resulted in the increasing use of expert opinions. These experts commonly express the site suitability in an ordered classification (e.g., unsuitable, slightly suitable, moderately suitable, suitable, very suitable) and are a relatively inexpensive source of information that can cover a wide variability of the landscape in a short period. The Global Assessment of Human-Induced Soil Degradation (Oldeman et al., 1991) based on qualitative expert assessments was a leading information source for land use policy-making, up to the present time.

Random sample. The compilation of the dependent variable follows a well-articulated geocoded strategy. We superimpose the map that indicates the available area and use a randomization protocol, available in most GIS software that respects a minimal distance between the randomly distributed grid cells. In case the GIS has no randomization protocols, the grid cells are read into a vector format and a random number is assigned to the cells to select a predetermined sample.
Field visit. Next, a field trip is organized to visit the randomly selected sites. For each site the square corresponding to the selected grid cell is determined as the assessment area. Next experts give their qualitative assessment on an ordered scale: very suitable, suitable, moderately suitable, slightly suitable, and unsuitable. The assessment is based on factors that are deemed important for the suitability of the site for allotment gardens.

For training purposes, the assessments can be compared and discussed among the experts in a plenary session.

Testing the experts. Compilation of expert judgments is seemingly a cheap and rapid way to collect data. However, expert opinions are also much disputed because they are not tested for consistency, while their quantitative interpretation is inherently unidentifiable. We address these identified weaknesses as follows. The consistency of the experts is checked by means of a cross-comparison of the expert judgments for sites with identical determinants (land use, biophysical characteristics, and socioeconomic and health conditions). As an example, we present the expert judgments given for a suitability assessment of sites for erosion hazard (Sonneveld, 2003). The findings are registered in a table that is presented in Fig. 9.5. The table can be read as follows. Following the third line below the column header, for example, we find in the first column the number of sites with identical characteristics, which is 4, and in the second column the total number that this group of identical sites occurs, here 6.

The third column indicates that in three cases the experts gave the same classification for all four sites and the fourth column shows that three cases reported one or more class differences. The fifth column shows how cases with different assessments are distributed: one case had three similar expert assessments and one

Number of identical sites	Occurrence of identical site group	Cases with equal expert judgement	Cases with different expert judgement	Cases with different assessment groups of identical sites with the same expert classification
2	47	24	23	23 1 + 1
3	15	3	12	10 2 + 1 and 2 1 + 1 + 1
4	6	3	3	1 3 + 1 and 2 2 + 1 + 1
5	3	1	2	1 4 + 1 and 1 2 + 2 + 1
6	11	3	8	1 5 + 1 and 1 4 + 2 and 1 4 + 1 + 1 and 2 3 + 3 and 3 3 + 2 + 1

Fig. 9.5 Table comparing expert opinions on site suitability concerning erosion hazards

assessment that was different, while in two cases three different expert judgments were given; two similar and two other, different assessments.

A quantitative interpretation of expert judgments. This section relates the qualitative expert assessments to a quantitative indicator that reflects the suitability of the site (e.g., yield of a standard crop). A straightforward approach is to use the ranges and average (or median) values of yields that correspond to the designated qualitative classes. A second approach is to estimate the quantitative values of class boundaries of a qualitative response model under the condition of a maximum likelihood of correct classification for all observations. We use an ordered qualitative response model (e.g., the ordered logit model), with the suitability classes of experts as the dependent variable while real-valued yields of the corresponding sites constitute the independent variable. The yield value that corresponds to the cutoff points of the classes is calculated from the estimated boundary values that, by default, are equal to the cumulative probability value of 0.5. These boundaries are scaled to yield measurements, expressed in ton per hectare, by dividing their value by the parameter estimate. Hence, we can define: $\frac{1}{1+e^{-(\mu_i-\beta X)}}=0.5$ and $x_{\mu_i}=\frac{\mu_i}{\beta}$, where x_{μ_i} expresses threshold values μ_i in physical units of the site realized physical yields.

Step 3. Preparation for Estimation: Conversion of Maps into Vector

With the dependent and independent variables compiled, we can start now with the preparation of the estimation steps. Therefore, we convert the maps into a vector format that is suitable for estimation exercises. Figure 9.6 gives an example of how the data are organized when a raster map is read as a vector.

For each grid cell defined by its coordinates, an attribute value is read and written in a vector format including its latitude and longitude.

Hence, for the maps carrying the independent variables (Fig. 9.7) we can convert each map into a vector format that leads to a matrix as shown in Fig. 9.8 where attributes of similar latitude and longitude are written as a row (Fig. 9.8).

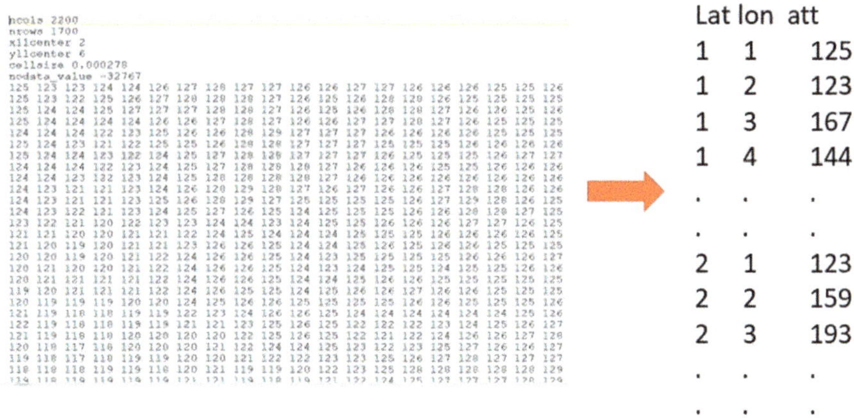

Fig. 9.6 Converting a map into vector format

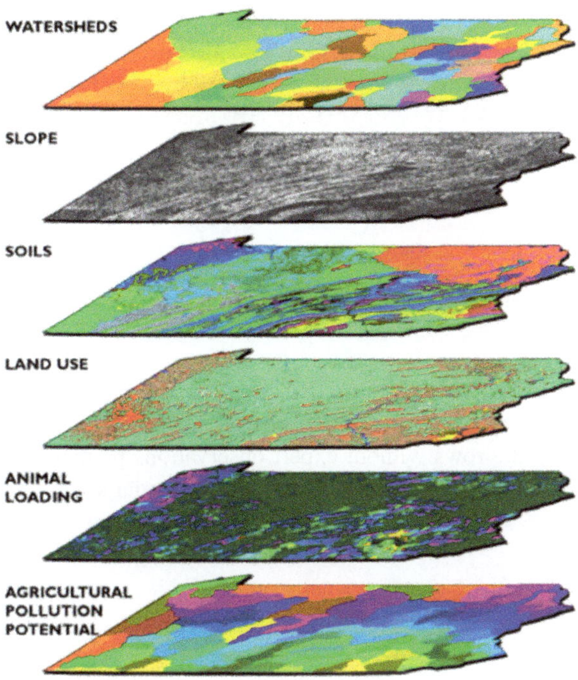

Fig. 9.7 Crossing of independent variables

Obs	lat	lon	soilsui	water	hours travel	distance markets
1	3	13	57	310	0.46	1767
2	4	15	31	240	0.46	2762
3	4	16	21	330	0.21	2732
4	4	17	38	390	0.21	2733
5	4	18	34	290	0.21	1666
6	5	14	39	190	0.23	6660
7	5	15	38	260	0.59	6420
8	5	16	6	130	0.59	5458
9	5	17	18	60	0.67	6454
10	5	18	30	130	0.67	4699
11	5	19	34	120	0.42	4699
12	5	20	34	100	0.42	4769
13	5	21	10	240	0.47	5760
14	5	23	16	140	0.45	5760
15	5	24	8	160	0.46	1757
16	6	14	35	100	0.76	1737
17	6	15	29	320	0.21	1637
18	6	16	21	320	0.21	2663
19	6	17	7	240	0.21	2663
20	6	18	20	330	0.23	5425
21	6	19	26	310	0.59	5425
22	6	20	30	60	0.63	6727
23	6	21	43	90	0.63	6684
24	6	22	33	30	0.63	4684
25	6	23	36	50	0.42	4704
26	6	24	22	170	0.45	4749
27	6	25	26	180	0.45	4716
28	7	13	24	200	0.69	2751
29	7	14	9	150	0.76	2731
30	7	15	14	170	0.76	1657

Fig. 9.8 Matrix format of independent variables georeferenced by the latitude and longitude coordinates

Finally, to prepare for the estimation protocol we append the corresponding expert observations that were given for selected sites creating a matrix form as in Fig. 9.9. Deleting the rows without expert observations prepares a full matrix for estimation of marginal contributions of independent variables.

Step 4. Estimation Protocol

The relationship between expert assessments of site suitability and a set of explanatory variables is formalized by estimating a qualitative response model. Site suitability for allotment gardens is represented by an unknown continuous variable y and related to independent variables x. Assuming additive error terms, the underlying process is represented as:

$$y_i = \beta' x_i + \varepsilon_i, \tag{9.1}$$

lat	lon	soilsui	water	hours travel	distance markets	expert observations
3	13	57	310	0.46	1767	
4	15	31	240	0.46	2762	.
4	16	21	330	0.21	2732	.
4	17	38	390	0.21	2733	.
4	18	34	290	0.21	1666	.
5	14	39	190	0.23	6660	.
5	15	38	260	0.59	6420	.
5	16	6	130	0.59	5458	.
5	17	18	60	0.67	6454	5
5	18	30	130	0.67	4699	.
5	19	34	120	0.42	4699	.
5	20	34	100	0.42	4769	.
5	21	10	240	0.47	5760	.
5	23	16	140	0.45	5760	.
5	24	8	160	0.46	1757	5
6	14	35	100	0.76	1737	.
6	15	29	320	0.21	1637	3
6	16	21	320	0.21	2663	.
6	17	7	240	0.21	2663	.
6	18	20	330	0.23	5425	.
6	19	26	310	0.59	5425	2
6	20	30	60	0.63	6727	.
6	21	43	90	0.63	6684	
6	22	33	30	0.63	4684	.
6	23	36	50	0.42	4704	2
6	24	22	170	0.45	4749	.
6	25	26	180	0.45	4716	.
7	13	24	200	0.69	2751	4
7	14	9	150	0.76	2731	.
7	15	14	170	0.76	1657	.

Fig. 9.9 Matrix form with independent and dependent variables

where i are observations, β the parameters to be estimated, and ε_i the disturbance, independent across observations. Ordered site suitability classes are observed as z_i. Adjacent intervals of y_i correspond with qualitative information z_i, as follows:

$$\begin{aligned} z_i &= 1 \quad \text{if } y_i < \mu_1, \\ z_i &= 2 \quad \text{if } \mu_1 \le y_i < \mu_2, \\ &\vdots \\ z_i &= n \quad \text{if } \mu_{n\ 1} \le y_i. \end{aligned} \tag{9.2}$$

Disturbances ε_i of the probability models are assumed to possess a logistic distribution, leading to a cumulative logistic transformation function Λ, with explicit form: $\frac{1}{1+e^{-(\mu_i-\beta X)}}$. Parameters β and thresholds ($\mu_1,\ldots, \mu_{n-1}$) are simultaneously estimated by maximizing probability of a correct classification by experts. Significance of estimated parameters is tested by Chi-square statistics while a likelihood ratio tests overall quality of the estimation.

Model accuracy. The estimated model is evaluated for its accuracy by a hit ratio, the percentage of correctly predicted observations. An example is given in Table 9.1 where the class predicted by a water erosion model is compared with the expert

Table 9.1 Cross-tabulation of expert assessment and model results

Cell frequencies (percentage of total)		Model estimate			
		No erosion	Slight	Moderate	Total
Expert	No erosion	107 (36)	33 (11)	0 (0)	140 (47)
	Slight	51 (17)	71 (24)	6 (2)	128 (43)
	Moderate	2 (2)	28 (9)	3 (1)	33 (11)
	Total	160 (53)	132 (44)	9 (3)	301 (100)

Source: Sonneveld & Albersen (1999)

assessment. It appears that the model reproduces 60% of the observations made by the experts. For 13% of the observations, the model gives a higher classification while, more seriously, for 27% of the observations, a lower class is predicted. It is also remarkable that the model predicts the "no erosion" class correctly for 76% of the observations. For the "slight erosion" class, this score is 55%, while the "moderate erosion" class scores a rather poor 9%.

Robustness. A tenfold cross-validation verifies model robustness for the inclusion or exclusion of observations. In this procedure, the data set is subdivided, at random, into ten sets of about equal size. The model is estimated each time with nine subsets of the data, keeping the remaining 10% as an evaluation set. The estimated parameters are applied to this evaluation set to compare model results with expert observations. In this way, ten different parameter estimates are obtained, as well as the site suitability class estimates of the entire set.

Figure 9.10 shows the results of a tenfold cross-validation conducted on an estimated site allocation tool (course material; NUFFIC). Most parameters exhibit minor fluctuations. Only the soil suitability is relatively sensitive. In this case, the results are sufficient to conclude that the estimation is not sensitive to the exclusion of observations.

Step 5. Decision Support Tool: Assess Suitability on Unvisited Sites for Various Options

We assess the suitability of unvisited sites (Fig. 9.11), where no assessment was made, as follows. First, estimated model parameters weigh the independent variables to calculate the probability of the related classes. The class with the highest probability was assigned to the unvisited sites.

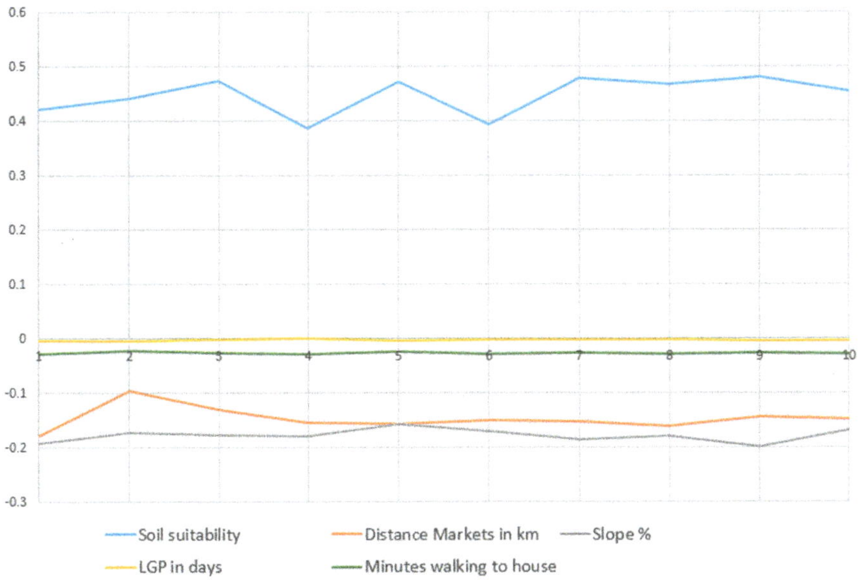

Fig. 9.10 Tenfold cross validation for site allocation model in Benin

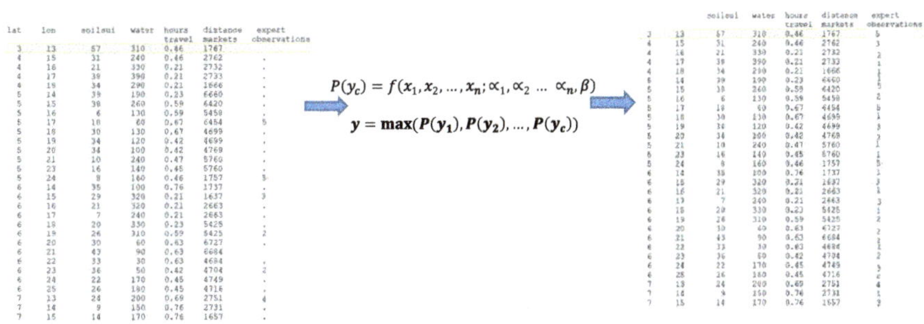

Fig. 9.11 Assessing site suitability at unvisited sites

Figure 9.12 shows how the estimated model $f(x)$, which applies the values of ∝'s to calculate the probability (P) of class c for all sites that were not assessed yet.

Step 6. Visualize Results: Conversion from Vector into

Dedicated software is written to read the matrix and place the selected attributes in an *.asc file format that is readable by most GIS programs. The principle is to place the attributes within the full matrix that is defined by the rows and columns of the map. Below a SAS program is placed that prepares a map (map.asc) and starts with placing the full map header (ncols, nrows, xllcorner yllcorner, cellsize

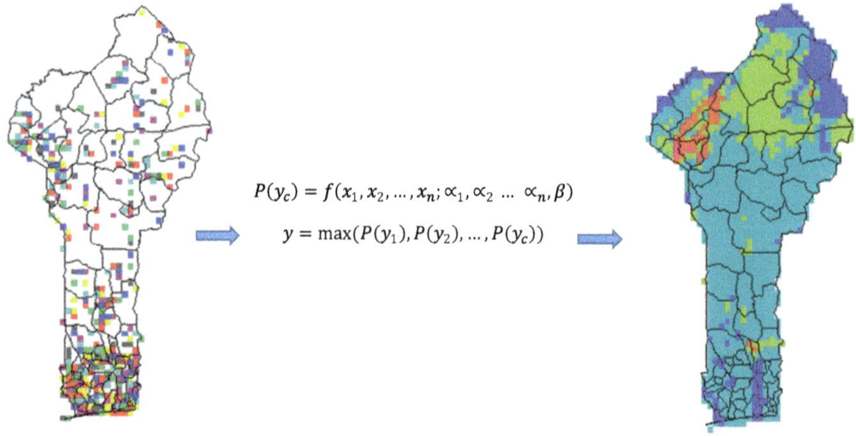

Fig. 9.12 Mapping unvisited sites

		soilsui	water	hours travel	distance markets	expert observations
3	13	57	310	0.46	1767	5
4	15	31	240	0.46	2762	3
4	16	21	330	0.21	2732	2
4	17	38	390	0.21	2733	1
4	18	34	290	0.21	1666	1
5	14	39	190	0.23	6660	1
5	15	38	260	0.59	6420	5
5	16	6	130	0.59	5458	2
5	17	18	60	0.67	6454	5
5	18	30	130	0.67	4699	1
5	19	34	120	0.42	4699	3
5	20	34	100	0.42	4769	2
5	21	10	240	0.47	5760	1
5	23	16	140	0.45	5760	1
5	24	8	160	0.46	1757	5
6	14	35	100	0.76	1737	1
6	15	29	320	0.21	1637	3
6	16	21	320	0.21	2663	1
6	17	7	240	0.21	2663	3
6	18	20	330	0.23	5425	1
6	19	26	310	0.59	5425	2
6	20	30	60	0.63	6727	2
6	21	43	90	0.63	6684	1
6	22	33	30	0.63	4684	1
6	23	36	50	0.42	4704	2
6	24	22	170	0.45	4749	3
6	25	26	180	0.45	4716	2
7	13	24	200	0.69	2751	4
7	14	9	150	0.76	2731	1
7	15	14	170	0.76	1657	3

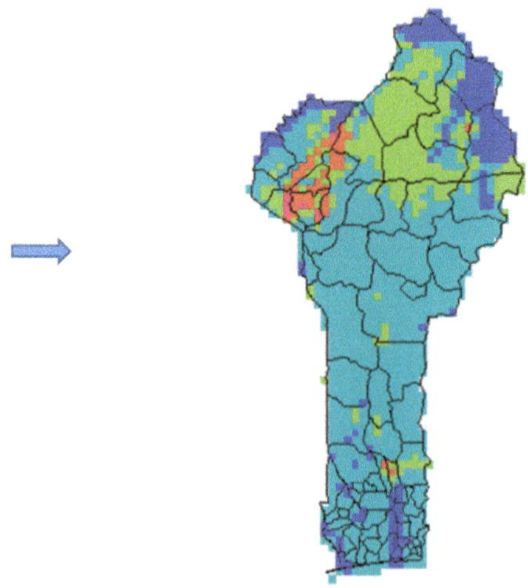

Fig. 9.13 From vector to map

nodata_value) that belongs to the asci format. Next it makes a loop that follows the number of columns (900) and places the value of the attribute (suitability) until the last column and then starts anew (@) in the next line. The resulting format can be depicted as a map (Fig. 9.13).

```
filename mapout "C:\ map.asc" ;
        data _null_ ;
          file mapout notitles lrecl=8239;
          if _N_ = 1 then do ;
            put "ncols 990";
            put "nrows 690";
            put "xllcorner -5.6292";
            put "yllcorner 9.3712";
            put "cellsize 0.00833";
            put "nodata_value -32767";
          end ;
          do i=1 to 990 ;
            set fullmap_inv ;
            put suitability @ ;
          end ;
          put ;
        run ;
```

Step 7. Policy Interpretation of Outcomes

In the last step we convert the estimated model into a site selection tool that can inform decision makers on:

- Potential candidate sites for allocation of allotment gardens.
- The impact of policy interventions on the suitability of the sites.
- In a baseline or reference scenario, we depict suitability of the sites without implementing discernible policies. Specifically, for all missing expert observations we calculate site suitability based on the estimated parameters, as explained in step 5.

The impact of policy interventions is done by manipulating the set of independent variables individually or in tandem. Following the linear equation in the ordered logit model we could write

$$\tilde{y}_s = \beta_0 + \beta_1 \tilde{x}_{1s} + \beta_2 \tilde{x}_{2s} + \ldots + \beta_n x_{ns}$$

where

$$\tilde{x}_{1s} = x_{1s} + \Delta x_{1s} \text{ and } \tilde{x}_{2s} = x_{2s} + \Delta x_{2s}$$

where, shock Δ alters the state of factors x_{1s} and $x_{2s,}$ at site s. For example, a fertilizer program that changes the current soil fertility status of site s will impact with a factor β_1 on site suitability y_s. Another example is to widen the distance to high population densities to avoid Covid-19 infections.

By manipulating the various independent factors one could design and evaluate the impact of various scenarios with respect to the base line: $y_s \sim \tilde{y}_s$. As noted, the scenarios can be made location dependent, for example, by selecting a subset of all districts, or soils with a certain fertility class. Using step 5 to calculate the results of the prospective scenarios and step 6 to visualize the results provides valuable information on the impact of the interventions.

Linking the results to a cost-benefit analysis (CBA) we can compare revenues and costs of a project in monetary terms using a discounting factor that expresses the results at current net present values (NPV):

$$\text{NPV} = \sum_{t=0}^{T} \frac{b_t - c_t}{(1+r)^t}$$

where T is the time horizon, b benefits, c costs, and r the discount rate. The information on CBA is also used to calculate the benefit cost ratio (BCR) by dividing total discounted benefits by total discounted costs, hence, estimating benefits for every monetary unit of investment.

Concluding Remarks

We presented seven fundamental steps to develop a site allocation tool that appraises the site suitability for allotment gardens in urban and peri-urban areas. The tool has a flexible format and its determinants can account for Covid-19 mobility restrictions and avoid high population concentration where prevailing infectious rates might be high. Houessou et al. (2021) clearly showed the benefits of access to allotment gardens in times of Covid-19 and the tool can be helpful to find the right sites while avoiding infection hazards and circumventing barriers and "cordon sanitaire." Hence, the allocation tool can support the decision-making processes about the integration of UA into urban land-use planning to achieve sufficient production and consumption of food in tangible results for better delivery. The tool can be expanded with a comprehensive business model that appraises the economic, social, and ecological feasibility of allotment garden projects. The tool supports city administrations to seek synergies in ensuring, inter alia, nutritious food, quality green spaces, prevention of heat islands, and buffer zones to absorb excess water. However, the development of a site allocation tool alone is not enough to further develop urban agriculture. A strong political will is needed at the various administrative levels with clear mandates to supervise the allocation of land. Further integration into urban development plans will mainstream the place of urban agriculture in the prevailing poverty alleviation policies.

References

Agnes, C., Louvet, A., & Bonneau, C. (2016). *Agriculture Urbaine et Justice Sociale* (p. 19). INRA Institut National de la Recherche Agronomique.

Bayendi-Loudit, S., Ndoutoume Ndong, A., & Francis, F. (2017). Le maraîchage périurbain à Libreville et Owendo (Gabon): Pratiques culturales et durabilité. *Cahiers Agricultures, 26*, 45002.

FAO. (2021). FAOSTAT. In *Population data, year 2000–2050 [online]. Rome.* Retrieved November 27, 2021, from http://www.fao.org/faostat/en/#data

Ferreira, A. J. D., Guilherme, R. I. M. M., & Ferreira, C. S. S. (2018). Urban agriculture, a tool towards more resilient urban communities? *Current Opinion in Environmental Science & Health, 5*, 93–97.

Friesen, J., & Pelz, P. F. (2020). COVID-19 and slums: A pandemic highlights gaps in knowledge about urban poverty. *JMIR Public Health and Surveillance, 6*(3), e19578. Published 2020 Sep 4. https://doi.org/10.2196/19578

Houessou, M. D., Cassee, A., & Sonneveld, B. G. J. S. (2021). The effects of the COVID-19 pandemic on food security in rural and urban settlements in Benin: Do allotment gardens soften the blow? *Sustainability, 13*, 7313. https://doi.org/10.3390/su13137313

Langemeyer, J., Madrid-Lopez, C., Mendoza Beltran, A., & Villalba Mendez, G. (2021). Urban agriculture—A necessary pathway towards urban resilience and global sustainability? *Landscape and Urban Planning, 210*, 2021. https://doi.org/10.1016/j.landurbplan.2021.104055

Lee, I. (2021). Urban gardening helping alleviate climate weight of cities 17 May, 2021. EuroScientist. Retrieved November 29, 2021, from https://www.euroscientist.com/urban-gardening-climate-weight-of-cities/

Musibono, D., Biey, E., Kisangala, M., Nsimanda, C., Munzundu, B., Kekolemba, V., & Palus, J. (2011). Agriculture urbaine comme réponse au chômage à Kinshasa, République Démocratique du Congo. *Vertigo-La Rev. Électron. Sci. L'environ., 11*. Retrieved March 18, 2021, from https://journals.openedition.org/vertigo/10818

Oldeman, L. R., Hakkeling, R. T. A., & Sombroek, W. G. (1991). *World map of the status of human induced soil degradation.* ISRIC/UNEP.

Pongutta, S., Kantamaturapoj, K., Phakdeesettakun, K., Phonsuk, P. (2021). The social impact of the COVID-19 outbreak on urban slums and the response of civil society organisations: A case study in Bangkok, Thailand, Heliyon, Volume 7, Issue 5, 2021, e07161, ISSN 2405–8440, https://doi.org/10.1016/j.heliyon.2021.e07161.

Rezai, G., Shamsudin, M. N., & Mohamed, Z. (2016). Urban agriculture: A way forward to food and nutrition security in Malaysia. Procedia Soc. *Behavioral Science, 216*, 39–45.

Rocklöv, J., & Sjödin, H. (2020). High population densities catalyse the spread of COVID-19. *Journal of Travel Medicine, 27*, 3. https://doi.org/10.1093/jtm/taaa038

Sonneveld, B. G. J. S. (2003). Formalizing the use of expert judgements for land degradation assessment: A case study for Ethiopia. *Land Degradation and Development., 14*, 347–361. Published online in Wiley InterScience (www.interscience.wiley.com). https://doi.org/10.1002/ldr.564

Sonneveld, B. G. J. S., & Albersen, P. A. (1999). Water erosion assessment based on expert knowledge and limited information using an ordered logit model. *Journal of Soil and Water Conservation, 54*(3), 592–597.

Verheye, W., Dent, D. (1997). Knowledge of the land. The International Land Use Society. Through: http://www.nri.org/ILUS/homepage.htm#1.1policy

World Bank. (2021). Population living in slums (% of urban population). In *World development indicators [online]. Washington, DC.* Retrieved November 28, 2021, from https://data.worldbank.org/indicator/EN.POP.SLUM.UR.ZS

Chapter 10
Tackling Ecological Overshoot: The Food System's Ten "Impossible Imperatives"

Mathis Wackernagel, Marta Antonelli, and David Lin

Abstract Global Footprint Network has identified ten "impossible imperatives" that shape the pressing challenges facing the global food system. These imperatives emphasize the simultaneous pressures the food systems are experiencing: the need to eliminate fossil fuel dependence while maintaining healthy and adequate food production for a still growing and increasingly urbanized population. Strategies include ensuring food equity, avoiding environmental degradation, and enhancing resource efficiency. Additionally, the food system must adapt to climate change, and embrace inevitable technological shifts, in a way that is economically viable. Given climate pressures, agriculture may also have to shift from being a greenhouse gas emitter to a sink. The most daunting challenge is addressing all these imperatives simultaneously, as partial solutions risk worsening other aspects of the crisis. Despite their seemingly impossible nature, the imperatives can be tackled with concerted effort. These challenges are not just about food security but are an essential element of humanity's broader transition toward sustainability.

Keywords Climate change · Global food system · Global Footprint Network · Impossible imperatives

Global Footprint Network and its partners[1] characterize the food dilemma through a set of ten tough challenges we call "impossible imperatives."[2]

[1] www.footprintnetwork.org/Food4Future

[2] A slightly earlier version was originally published as a blog at www.footprintnetwork.org/2021/10/18/tackling-ecological-overshoot-the-food-systems-10-impossible-imperatives

M. Wackernagel (✉) · M. Antonelli · D. Lin
Global Footprint Network – California, Oakland, CA, USA
e-mail: mathis@footprintnetwork.org

A. A. Kubursi et al. (eds.), *Urban Food Security in a Crisis Prone World*,
https://doi.org/10.1007/978-3-031-89440-4_10

The Imperative to Use No More Fossil Fuels

The food system, and particularly farms, will have to produce all the food while rapidly phasing out fossil fuels or its derivatives from farm to fork. Humanity, if it wants to comply with the Paris Agreement and avoid the worst of climate change, will have to get out of fossil fuel use well before 2050. Even if humanity does not want to, eventually it will be forced to quit its fossil fuel dependence, and will face even more unpredictable climate extremes, if it were to phase out those fuels slowly.

The global food system is heavily fossil fuel dependent, with each food calorie requiring several fossil fuel calories to be produced.[3] Currently, fossil fuel is a cheap and convenient enabler of the food system. Yet these strengths come with serious downfalls. These include but are not restricted to depleting farm practices starting with excessive fertilizer use, ability to pumping massive amounts of irrigation water, and operating intricate machines. Also, many fossil fuel derivatives, such as plastics, are used widely from field to fork. This imperative is challenged even more by the need to part ways with fossil fuels without losing current levels of crop yields.

The Imperative to Increase Food Production and Reduce Food Waste

There is an increasing demand for food from a growing population that is increasingly urbanized. The demand is getting amplified by shifting diets: many people seem to want ever more protein in their diet, especially animal-based protein—even though they may not have a protein shortage in their diet. Typically, protein calories take far more bio-capacity than general food calories (such as starches), even if the protein is from plants rather than animals. Some of the increased food demand can be provided by cutting food loss and waste. Currently, estimates suggest that approximately one-third of food produced is wasted. Also, the increase in demand, which may well exceed the growth rates of populations, may be so high that there will be a need for higher yields. Some may argue though that increases in population size, even globally, may not need to occur. But surprisingly, there is little consensus among political leaders, that smaller populations would be of benefit to each region, country, or even humanity as a whole. Without such support, the demographic transition may continue to be rather slow. (Explore your demographic assumptions using our downloadable scenario calculator.)[4] Therefore, higher future demand is likely.

[3] Vaclav Smil, 2022, How the World Really Works: The Science Behind How We Got Here and Where We're Going, Viking, NY.

[4] The scenario tool can be found here: www.footprintnetwork.org/resources/footprint-scenario-tool

The Imperative to Contribute to Food Equity

Food for all is of paramount importance, given the largely uneven distribution of economic possibilities, even though high-income areas are generally well supplied with food and struggle more with avoiding obesity among their populations, rather than food scarcity. Starvation is still a real threat in many parts of the world. An estimated 3 billion people do not have access to a healthy diet, according to FAO, and diet-related noncommunicable diseases are on the rise and cause disability and mortality. Given climate change, growing demand, and the need to move out of fossil fuel, the food security situation could get tighter again. Today, 72% of humanity lives in countries that run an ecological deficit and live on an income smaller than the average global income per person.[5] This means 72% of the world population will be at a particularly brutal disadvantage competing in international markets for resources and food once ecological deficit spending will become less of an option. Just on food alone, over 3 billion people live in a country where domestic food biocapacity is below the amount of food the country's population consumes and with a lower than global average income per person.[6] They are most immediately exposed to food shocks.

The Imperative to Avoid Degradation and Pollution

Farms will have to operate in ways that do not erode or contaminate their soils and waterways, now and in the future. Given our fragile planet, humanity can no longer afford to degrade or lose soil, overexploit and deplete surface and groundwater, assault biodiversity, or apply substances emitting persistent pollutants. There may even be calls for farms to become regenerative, agro-ecological, or even restorative.

The Imperative to Become More Resource Efficient beyond Fossil Fuel Elimination

It is not just fossil fuel input that will become ever less available. Resource scarcity reaches across the entire pallet of material inputs. It is no longer reasonable to count on being able to access more water, particularly groundwater, more wood for structures and fences, more cotton for shading or packaging, etc.

---

[5] Wackernagel, M., Hanscom, L., Jayasinghe, P., Lin, D., Murthy, A., Neill, E., Raven, P., 2021. The importance of resource security for poverty eradication. Nature Sustainability. Volume 4, pages 731–738. https://doi.org/10.1038/s41893-021-00708-4

[6] The calculation is based on the National Footprint and Biocapacity Accounts, now maintained by York University in Toronto (https://data.footprintnetwork.org and https://info.footprint.yorku.ca). The idea is elaborated more on the blog available here: https://www.footprintnetwork.org/2022/07/13/food-security-in-a-world-of-overshoot/

The Imperative to Provide Greenhouse Gas (GHG) Sinks

To live up to IPCC scenarios, agricultural land must become a sink of GHG, not a source as it is now, particularly post-2050. This means that farms will have to change practices that emit additional greenhouse gases, particularly methane. Such GHGs may come from land conversions, cattle, or rice paddies. Since IPCC scenarios all bank on quite massive net sequestration in the second half of the twenty-first century, just eliminating GHGs from agricultural production may not be sufficient. While there are some promising examples of how this can be achieved, nobody knows yet how to do this at the appropriate scale without diminishing food production.

The Imperative to Prepare the Food System for Significant Climate Shifts

Because we can no longer expect and depend on the usual, regular weather patterns, the food system must become climate resilient. Extremes will be more common, overlaid, most likely, with an overarching warming trend (at least as a global average). The average climate may be heating up more, or less, in your own region, with increased seasonal variance. This may affect the evapotranspiration conditions of what you grow. Therefore, it may require significant adaptation efforts, possibly growing different kinds of crops, shifting location of production to the urban areas, and adjusting the contours of your territories to cope with changing water availability. Climate change mitigation will also be needed, with urban food systems playing a key role as almost 70% of global population is assumed to be living in cities by 2050.[7]

The Imperative to Navigate the Inevitable Technological Shifts

Significant technological shifts are increasingly likely, and probably escape attempts to regulate those technologies' use. Whether AI (artificial intelligence) is taking over farm management, robots are working the fields, or supply chains are automated,

[7] https://www.un.org/development/desa/en/news/population/2018-revision-of-world-urbanization-prospects.html

many such changes are coming, with both challenges (such as social inequity and ecological impacts) and opportunities. As a consequence, the food system professionals will have to learn new things rapidly. Most of these things are not yet taught at schools, and some of those professionals might even get displaced by emerging technologies. Society may be able to delay adoption of such technologies, but delays may also risk losing out on productivity gains. Science and technology will need to back up local traditional knowledge.

The Imperative to Succeed Even in the Absence of Financial Upsides

Tragically, in spite of massive ecological overshoot, the stewards of biological capital are rewarded poorly. For instance, low-income countries with high share in agriculture do not seem to benefit financially from their access to higher levels of bio-capacity (see Fig. 10.1). One reason is that the benefits from value chains, all of which depend on bio-capacity, are not distributed back to the stewards of natural capital. Most benefits of the value chain go to the brands and urban distributors.

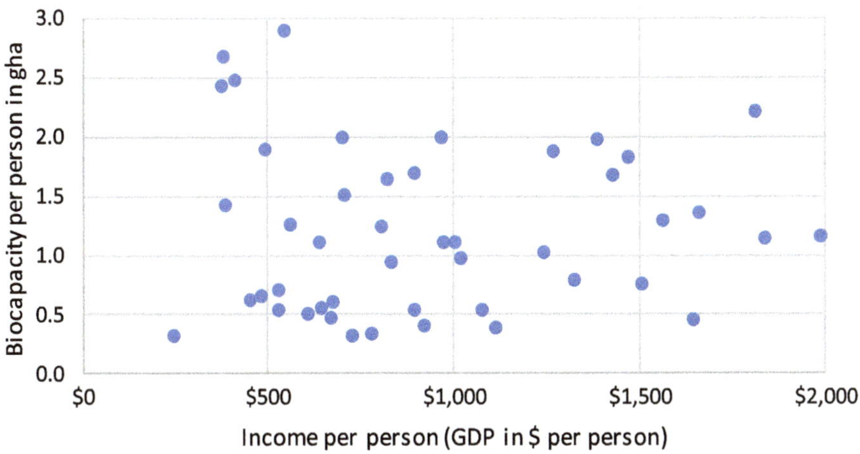

Fig. 10.1 National biocapacity and income. ***Notes:*** *This graph shows, for countries, per person biocapacity as well as per person income. The figure shows that reality contradicts our initial assumption that more biocapacity per person (as a proxy for more farm capacity per person) would lead to higher incomes, even among countries with below 2000 USD annual per person income where farming is a dominant contributor to GDP. For these countries, there was no, or even a slightly negative, correlation between biocapacity per person and income (2014 data). Note that one country with a biocapacity of 7.4 gha per person, the Central African Republic, is missing from this graph, and would, with a GDP of around 300 USD per person be at the lower end of the scale.* (Source: From Wackernagel, M., Lin, D., Evans, M., Hanscom, L., Raven, P., 2019. Defying the Footprint Oracle: Implications of Country Resource Trends. Sustainability, Volume 11 Issue 7, 2164, https://doi.org/10.3390/su11072164)

Even if farm production increases, this does not seem to boost farm income as larger production can decrease prices. Urban markets do not value the food system sufficiently, particularly those at the foundation of the value chains.

The Imperative to Address Them All in the Absence of Societal Support

All of these imperatives are hitting the food system at once. It is not about picking and choosing from the nine problems mentioned above. To succeed, humanity needs to address all of them simultaneously, and not any at the cost of any other one. Otherwise, humanity will not be able to solve its food problem. And yet, these massive challenges are not yet recognized as a societal priority. Hence, the Global Footprint Network and its partners' focus on exploring the implications of these challenges and to contribute to identifying pathways.

Concluding Remarks

Obviously, calling these ten conditions "impossible imperatives" is merely a humorous provocation, as we believe that they can be overcome, if tackled vigorously. But we also recognize that all of them need to be overcome together. This may prove to be as challenging, if not more, than the energy transition needed to address climate change. Based on these imperatives, we are exploring with our partners how they can be overcome as a whole. There is no doubt that the food system needs overhaul, from farm to fork. Hence, we at Global Footprint Network are always looking for partners interested in making this transition a reality.

Chapter 11
COVID-19 and Food Security in MENA Countries

Dina A. Mandour

Abstract We examine the link between the pandemic and food security in the Middle East North Africa (MENA) region, highlighting the various channels through which COVID-19 impacts key pillars of food security: affordability, availability, and utilization. By estimating two equations with distinct indicators for food security and proxies for the effect of the pandemic, the study finds that higher confirmed COVID-19 cases and weaker national preparedness to handle the disease are significantly associated with increased food insecurity levels. The empirical assessment underscores the relative importance of the role of institutional and demographic prerequisites needed to handle the pandemic in explaining food insecurity variability across all countries, suggesting that these factors are more influential than the pandemic's stringency per se. The findings indicate that the MENA region faces specific challenges that put it in a disadvantaged situation compared to other regions, including weak governance, high levels of corruption, and fragile health systems, making it particularly vulnerable. Consequently, the pandemic may act as a catalyst that would intensify the urgency to undertake radical reforms in food systems and to revisit several structural and institutional rigidities that have affected accessibility and utilization pillars in the MENA region.

Keywords Food security · Food insecurity · COVID-19 · Pandemic · Middle East North Africa

D. A. Mandour (✉)
Faculty of Economics and Political Science, Cairo University, Giza, Egypt

A. A. Kubursi et al. (eds.), *Urban Food Security in a Crisis Prone World*,
https://doi.org/10.1007/978-3-031-89440-4_11

Introduction

COVID-19 escalated from a public health emergency status to be declared by the WHO on the 11 March 2020 as a global pandemic. The pandemic has been described as "an extraordinary situation" where it represents a crisis of a unique nature that has affected all countries across the globe regardless of their economic status (ITC, 2021). Analogously, within each country, most of the segments and actors in the economy were negatively affected, though with different intensities. COVID-19 crisis is a multidimensional crisis, where at the outset it is a health crisis, but at its core it is an economic as well as food security (FS) crises, hence making the world featuring a triple shock (Erokhin and Gao, 2020; Mouloudj et al. 2020). The impact of COVID-19 on FS is still unfolding but as articulated by UN (United Nations UN 2020a, b), it could represent "an impending global food emergency of unknown, but likely very large proportions." FAO has declared that FS of around 820 million people will be hard hit by the crisis, with 62% and 31% of those affected concentrated in Asia and Africa, respectively (Oraikat, 2020). The most substantial increase in the number of people suffering acute hunger in 2020 is likely to be concentrated in least developed countries (LDCs) of Sub-Sharan Africa, and developing economies of Middle East North Africa (MENA) and Latin America (Erokhin and Gao, 2020). Some forecasts suggest the addition of 83–132 million people to the ranks of undernourished in 2020 due to COVID-19 (FAO et al., 2020). Moreover, with the shrinkage of global trade that can reach a decline of 32%, food supply chains would be highly disrupted (Petetin, 2020). The pandemic has posed considerable various challenges and risks in an array of areas and dimensions. Among such challenges is the momentum of Sustainable Development Goals (SDGs), which is likely to stagnate and many of the goals on top of the development agenda would be forced to recede. Furthermore, COVID-19 is highly expected to cancel out most of the previous and ongoing efforts and initiatives undertaken to improve FS status in many countries (Kinsey et al., 2020).

Combined with sharp resource reallocation to the health sector and the significant drop in income, it is highly expected to have inflated food insecurity (FI) rates across the globe and especially within those countries with initial alarming FI status. MENA region is among the highly vulnerable regions in the world together with Sub-Saharan Africa and Latin America being subject to suffer from acute hunger amid COVID-19 crisis (Erokhin and Gao, 2020). The MENA region being cursed with relative scarcity of natural resources, political conflicts, and macroeconomic instabilities has put the pre-pandemic FI status in the region on the edge. The pandemic is highly expected to further magnify the preexisting weaknesses in several MENA countries where the combination of weak institutions with disruptions to food supply chain and food aid will deepen the stringency of the crisis (FAO, 2020). The initial FI status in many countries in the region defines to a great extent the degree of vulnerability to such crisis that could hinder them from an optimal response to the crisis (Mandour, 2017). The macroeconomic performance and FS management signal another source of their fragility in facing COVID-19 crisis. Yet, as much as COVID-19 could potentially result in alarming risks and threats to FS rates, the picture is not totally gloomy, as so far global food availability has been adequate and there has been no worrisome price hikes.

This chapter aims at assessing the link between COVID-19 and FS status with special focus on MENA[1] countries in a trial to classify them according to the degree of risk they could be susceptible to. The study aims also at outlining the main transmission channels through which the crisis could affect FS in general and in the MENA region in particular. Within this context, the chapter would highlight the specific nature of MENA countries regarding the impact of the crisis on FS status and the factors behind variations of such impact. Finally, the study intends to quantitatively assess the impact of the pandemic on FS across countries with special emphasis on the relative situation of the MENA region.

The chapter is structured as follows. Following the introduction, *Section One* portrays the status of food security in the MENA region to comprehend the status of FS pre- and amid the COVID-19 crisis. *It* also outlines the various measures undertaken by MENA countries to mitigate the negative impacts of the pandemic. The quantitative analysis is undertaken in *Section Two* where a cross-sectional regression is applied to empirically investigate the nature of the relationship between the pandemic and the status of FS with special emphasis on the effect on MENA countries. *Section Three* concludes and highlights the policy implications.

Food Security in MENA and Potential Impact of COVID-19 Crisis

Status of Food Security and COVID in MENA

The MENA region holds the second rank after the Sub-Saharan countries with the highest prevalence of undernourishment (Fig. 11.1). A number of countries in the region are highly vulnerable and their status of FS has been worsening over time. The region as a whole is among the largest world food importers where on average each country imports more than 50% of the calories it consumes. The utilization is the weakest FS pillar in the region where around 52 million people suffer from obesity and other food-related diseases as diabetes (Karasapan, 2020). Moreover, the region includes two of the top ten countries in the world enduring highest severity of food crises amounting to 53% of the population in Yemen and 36% in Syria (GRFC, 2020). Along with Yemen and Syria; Iraq, Libya, and Palestine experience one of the alarming conflict-related humanitarian emergencies (FAO, 2020). This brings conflict and

[1] The paper follows the World Bank classification where the Middle East and North Africa (MENA) countries include: Jordan, Egypt, Syria, Qatar, Iran, Algeria, Bahrain, Djibouti, West Bank and Gaza, Iraq, Israel, Libya, Kuwait, Lebanon, Malta, Morocco, Oman, Saudi Arabia, Tunisia, United Arab Emirates, Yemen.

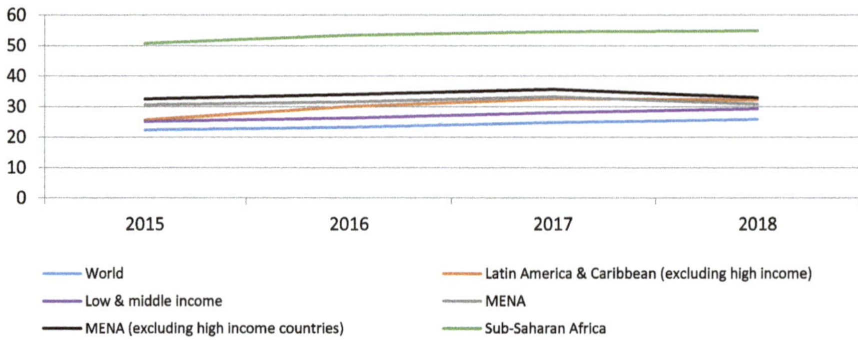

Fig. 11.1 Prevalence of undernourishment (PoU) (% of population). (Source: Adapted from WDI (2020))

instability as main triggers of FI in the region (GRFC, 2020). According to the Global Hunger Index (GHI), Yemen and Syria reached extreme alarming scores (35–49.9),[2] followed by Iraq (17.1) and Egypt (11.9) having 2 digit hunger index scores and classified as moderate degree of risk yet still high (see Table 11.2 in Annex). In other countries like Lebanon and Turkey as well as Jordan, refugees face severe degrees of FI (Fig. 11.2). Gulf Cooperation Council (GCC) countries are home for huge number of migrant workers, forming a considerable percentage of the population exceeding 80% in the United Arab Emirates (UAE) and 70% in Qatar and Kuwait, and 30% in Bahrain, Saudi Arabia, and Oman (Woertz, 2020). This substantial category of the population faces several degrees of discrimination where they are not covered by subsidy schemes as normal citizens and are deprived from basic labor rights. Above all, the living conditions of the majority hamper them from attaining the minimum requirement for nutritious food (Woertz, 2020). All this puts their FI status on the edge. Both segments of the population, migrant workers and refugees, "the invisible martyrs of the food system," as named by Zurayk (2020), are subject to alarming FI levels. Furthermore, Lebanon has been currently facing political and macroeconomic severe crises where the soaring public debt is placing strain on foreign currency reserves and leading to reduced capacity to import essential goods including food. Moreover, the depreciation of the currency by more than 63% from October 2019 to February 2020 has eroded the purchasing power of the population and moved a large segment of the population to poor and near-poor status (GRFC, 2020). The situation is also bad in other countries where 9% and 6% of the population in Iraq and Libya, respectively, require food assistance (Karasapan, 2020; Ling Ma et al., 2021).

The relatively limited arable land in the region is also a main problem factoring in and adding more complexity to the FS issue. Only 30% of the land in MENA is suitable for agriculture. Moreover, all MENA countries suffer from water stress ranging from high (40–80%) to extremely high (more than 80%) usage to supply

[2] https://www.globalhungerindex.org/ranking.html

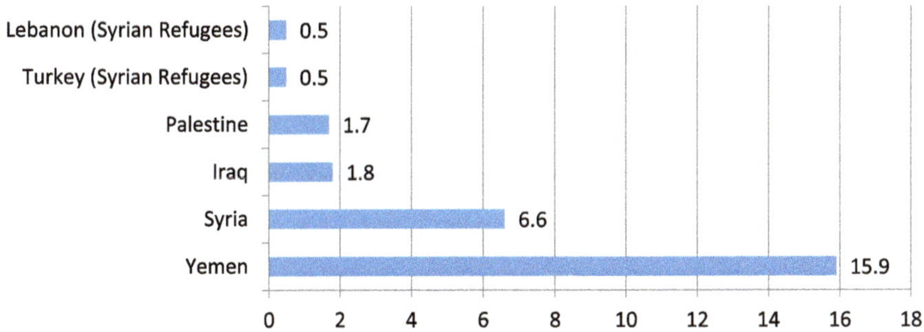

Fig. 11.2 Number of acutely food insecure people (million). (Source: GRFC (2020))

Table 11.1 Food insecurity in MENA region amid COVID-19 (excluding high-income countries)[a]

	People with insufficient food consumption (January 2021)	% of Population	People with insufficient food consumption (October 2020)
Algeria	↑5.6	13.3%	5.34
Djibouti	0.3	30.0%	0.38
Egypt	↑10.7	10.9%	10.64
Iran	6.2	7.6%	6.62
Iraq	2.9	7.6%	3.27
Jordan	1.2	12.0%	1.25
Lebanon	0.9	13.2%	0.94
Libya	↑0.7	10.4%	0.65
Morocco	↑5.7	15.8%	5.24
Tunisia	↑2.1	18.1%	1.81
Turkey	↑11.7	14.2%	10.91
Yemen	12.9	43.0%	13.01
Syria	9.8	48.0%	9.12
Palestine	1	21.7%	0.99

Source: extracted from WFP (2021)

[a]The table excludes GCC countries as well as Israel since they do not suffer from a significant undernourishment problem and thus are not covered by the World Hunger Map (WFP, 2021)

ratios (Karasapan, 2020; World Resources Institute, 2013). Such water poverty puts FS under significant stress. Moreover, the policies adopted by the majority of MENA governments have exacerbated the problem of FS where generous cheap water was provided and at the same time food consumer subsidies, promoting unhealthy diet, have been a main policy tool (Karasapan, 2020; Woertz, 2020).

Referring to the GFSI scores (Table 11.1), Syria and Yemen have remained on the tail of the list in the overall score as well as within each pillar. High-income countries in the region, GCC countries and Israel, score the highest in the affordability and utilization pillars (except for Bahrain) and relatively low in the availability dimension where Israel, Egypt, Turkey, and Morocco come as best performers in this pillar in the region. On the global level, Tunisia, Algeria, and Yemen were among the ten most deteriorating countries whereas Kuwait and Qatar were among the best ten improving countries worldwide.

With the prolongation of COVID-19 pandemic, FS is brought under extreme scrutiny in MENA. The number of people with insufficient food consumption increased in the short period between October 2020 and January 2021 in several MENA countries including Algeria, Egypt, Libya, Morocco, Tunisia, and Turkey (Table 11.1).

On another note, the pandemic is highly expected to cause disruptions to food supply chain and food aid (FAO, 2020). It is worth noting that food aid is vital for some countries where 50% of the population in Yemen and Syria is dependent on food aid (Karasapan, 2020). The highest fatality rates within COVID confirmed cases occurred in Yemen, Turkey, Egypt, and Syria (Fig. 11.3 and Table 11.2 in Annex). Fig. 11.4 shows a negative relationship, within the MENA region, between the countries' ability to handle the pandemic according to their demographic, economic, and institutional rigidities as expressed by the INFORM COVID-19 risk index and food security as measured by GFSI.

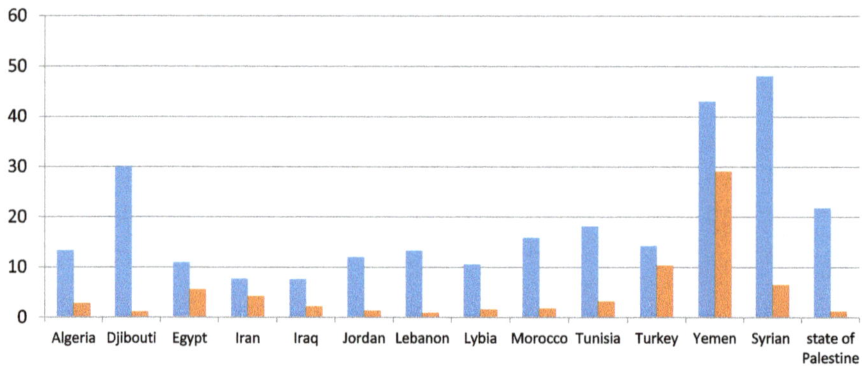

Fig. 11.3 Percentage of population with insufficient food consumption and Intensity of COVID as of January 2021. (Source: Adapted and calculated from World Food Program WFP (2021))

Table 11.2 Definition of variables in Eq. (11.1) and their data sources

Variable	Definition	Data source
GFSI (dependent variable)	Global FS Index	EIU (2020)[a]
GDPC	GDP per capita at constant prices (=2015)	UNCTADSTAT (2020)[b]
HAZ	COVID-19 Hazard and Exposure Index	Poljanšek et al. (2020)[c]
COP	Lack of coping capacity	Poljanšek et al. (2020)
DES	Average dietary energy supply adequacy (percent) (3-year average)	FAOSTAT(2020a)[d]
FDINF	Food inflation	FAOSTAT(2020a)[e]
RFDIMP	Ratio of food imports to total exports	Calculated from UNCTADSTAT (2020)
MENAHAZ	Dummy variable that takes a value of 1 if the country belongs to the MENA region and 0 otherwise along with its score in the hazard and exposure index	
MENACOP	Dummy variable that takes a value of 1 if the country belongs to the MENA region and 0 otherwise along with its score in lack of the coping capacity index	

[a]https://foodsecurityindex.eiu.com/

[b]https://unctadstat.unctad.org/wds/ReportFolders/reportFolders.aspx?sCS_referer=&sCS_ChosenLang=en

[c]https://drmkc.jrc.ec.europa.eu/inform-index/INFORM-Covid-19

[d]http://www.fao.org/faostat/en/#data/FS

[e]http://www.fao.org/faostat/en/#data

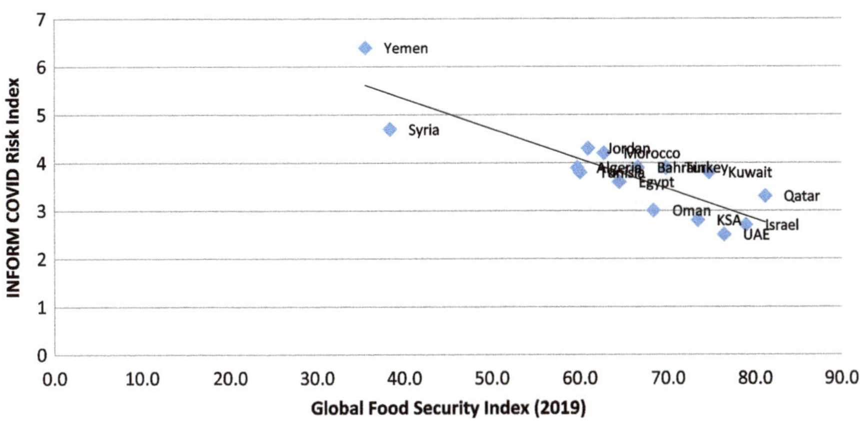

Fig. 11.4 COVID-19 Readiness Risk Index (INFORM COVID Risk Index) and Global Food Security Index in MENA countries. (Source: Adapted from Poljanšek et al. (2020); EIU (2020))

Channels of Transmission of COVID-19 Impact on FS in MENA Region

The pandemic has been generating a set of disruptions that would be transmitted to FS within the region via various channels.

Economic Downturn and Relative Scarcity of Natural Resources

There are direct channels as well as indirect transmission channels by which the COVID-19 crisis is likely to affect FS in MENA countries. Among the direct channels is the slowdown of the world economy and its expected repercussions. According to the World Bank Group (2020), a downslide in the economies of the region was expected to be by 5.2% in 2020. Such projections were constantly revised bringing up more pessimistic trends on the scene. According to World Bank (2021) projections, MENA's collective GDP was expected to contract further by 7.6% points in 2021. At the country level, the World Bank (2021) calculations registered the highest expected GDP losses in Lebanon, reaching a decline of 30% followed by Yemen and Iraq with a descend of about 15% then Kuwait, Oman, UAE, and Tunisia with an average of 9%. Algeria and Morocco follow by 8% while Egypt held the least downgrade in the region reaching 4.6%. The 2021 forecast for recovery was also dim where countries' GDP was expected to remain below their pre-crisis level (World Bank, 2021).

The twin shock of COVID and declining oil prices expected to hurt countries unevenly where the World Bank (2021) projections revealed that conflict countries and oil exporters are the most severely affected countries (Yemen, Iraq, Libya, and Syria), followed by the GCC and oil importers. Oil importers are those MENA countries suffering from large fiscal deficits and high cereal import dependency (e.g., Jordan and Lebanon) whereas the GCC countries are cushioned by a better fiscal position enabling them to absorb the increasing costs, albeit being fully dependent on wheat imports and facing simultaneously dwindling world oil prices for their major export good (Larson et al., 2012). The declining oil prices have negative potential impact on other variables including investment, remittances, and aid flows. For example, remittances were projected to drop by 20% in 2020. Further, in 2020 FDI to GCC declined by 20.2% whereas in non-GCC MENA countries it fell by 74% compared to 2019 (World Bank Group, 2020). This would indeed magnify the recessionary impact of the pandemic to an extreme level that would be transmitted to FS on the affordability as well as the utilization pillars.

Moreover, according to the World Bank (2020) the agricultural output in MENA is likely to be negatively affected (3% reduction) equivalent to the reduction in manufacturing, yet less severe than services (9.3%) (World Bank Group, 2020). The reduction in output is likely to imply reduction in employment where most of the agriculture in MENA countries is dominated by small-scale agricultural producers and informal sector. In addition, the demand side is negatively affected where the purchasing power of the poor and vulnerable segments is expected to decline further which will deepen the FS negative status.

Focusing on domestic production, prospects differ significantly among MENA countries. For example, drought has significantly affected Maghreb countries (Morocco, Algeria, and Tunisia). All three countries produce mostly rain-fed cereals. As a result of the drought wheat production in Morocco was expected to be reduced by 50% of the country's average in 2020. Yet, all three countries have adequate cereal stocks. In other countries, such as Syria and Iraq, weather conditions remain favorable; however the agricultural production is disrupted by the domestic conflict they have been experiencing for a relatively long time. The prospects for grain availability in the coming months are relatively good in Egypt, the world's largest importer of wheat, providing enough grain to meet national demand for several months (FAO, 2020).

Anecdotal evidence suggests that COVID-19 has significantly negatively affected the status of FS in a number of MENA countries. For example, World Bank Group (2020) shows that the Tunisian population suffered tremendously where 18% of the households consume less food and the poorest quintile reduced its food consumption fivefold amid the crisis despite the availability of a number of social protection programs. The same is true in the case of Libya, yet in a more drastic manner, where 70% of the households consume less food compared to the situation before the crisis. (Erokhin and Gao, 2020; World Bank Group, 2020). In Jordan, agriculture was most of the highly affected sectors by the crisis due to lockdown measures, which resulted in shortage of labor. Yet, the most significant loss was incurred by the food services sector (namely, hotels and restaurants) which is expected to suffer losses reaching over 90% (Raouf et al., 2020). Breisinger et al., (2020) found that the crisis's effect on FS in Egypt is significant. The negative impact is channeled mainly through the economic downslide affecting the food services sector, which is likely to suffer significant losses exceeding 50% and indirectly affecting the food processing industry, yet with less severe effect. However, the agriculture sector, as a component of agri-food system, remained resilient with an expected significant increase in output. Moreover, the agricultural sector is not likely to experience reduction in production for all crops, but is likely to experience an increase in wheat production due to higher government procurement prices (Breisinger et al., 2020).

Such disruptions have led to price increases in the MENA region, which have negatively affected the accessibility pillar of FS. Since February 2020 the prices of main food categories in the region have been on the rise including those for carbohydrates, diary, fruits, meats, and vegetables. Yet the price increases have been moderate ranging 5% on average for the majority of food products with an exception of food staples where price increases reached 20% and more since February 2020 in several countries including Djibouti, Egypt, Iran, Kuwait, Lebanon, Morocco, Qatar, Saudi Arabia, Syria, and Yemen (World Bank, 2021).

From another highly related perspective, poverty has been deeply rooted in the region with an estimated 20.3% in 2018 rising by 33% compared to 2015 (WDI, 2020). The general prospects for growth in the region remain dim where the IMF has made its lowest forecasts for the region in 50 years (UNSDG, 2020), hence deepening the poverty aspect. The current crisis is estimated to increase the poor in the region by 14.3 million to reach 115 million people by end of 2020 (UN, 2020a). The recent World Bank poverty projections due to COVID are even gloomier where the rate of increase in poverty is expected to range from 15% to 100% as depicted in Table 11.3 in Annex (World Bank, 2021). In such a situation, it is expected that the COVID-19 crisis is likely to exacerbate the already existing extremely vulnerable FS situation (FAO, 2020).

Political Instability

Among the three main pillars of FS, namely, two (availability and affordability of food) were always among the slogans that were raised during the different uprisings that started in 2010 onward, indicating the severity of the problem and its historical, political, and economic dimensions. Paradoxically, the existing regimes before the uprisings have had food subsidy programs enacted, however, they were not capable of ensuring FS in reasonable terms. Political instability has a direct link to FI in various dimensions. For example, encroachment of urbanization on agricultural land in Egypt intensified in the aftermath of January 2011 revolution making use of the weak subsequent governments, lack of security, and absence of rule of law. About 62,000 hectares of agricultural land was lost in the 3 years after the revolution, a two-third increase than the normal annual loss bringing up extreme pressures on FS (Mandour, 2017).

International Trade

Trade is a vital channel through which COVID can significantly affect FS. The fact of being highly dependent on food imports exacerbates the FI status in the whole region. The region is the largest importer of cereals globally, where GCC countries (excluding Saudi Arabia) import 100% of their cereal needs, and countries as Jordan, Oman, Palestine, Yemen, Saudi Arabia, Libya, and Lebanon import more than 90% of their grain needs (Fig. 11.5). Such high import ratio makes such countries prone to any distortion in world food prices, currency volatilities, and food availability shocks (Mouloudj et al., 2020).

A number of MENA countries depend highly on exporting a number of food products (Fig. 11.6). Such products constitute a significant share in countries like Lebanon, Palestine, Syria, Morocco, Egypt, Jordan, and Turkey (WDI, 2020). Depending on food exports, among which are highly perishable ones, imply more vulnerable exposure to supply chains and logistics' disruptions (FAO, 2020). Such countries are expected to face a further decline in their food exports following the

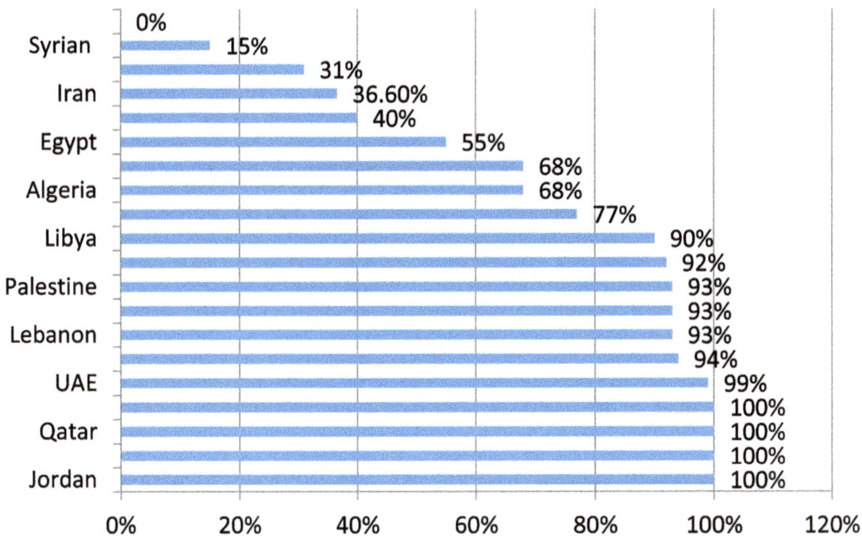

Fig. 11.5 Import dependency ratio (Percentage of a country's imported food for domestic supply versus its own food production for domestic supply; it is calculated as follows: IDR = Imports/(local production + imports—exports)*100% (WFP, 2021))(Cereal). (Sources: Adapted from WFP (2021); FAO (2020), and FAOSTAT (2020b))

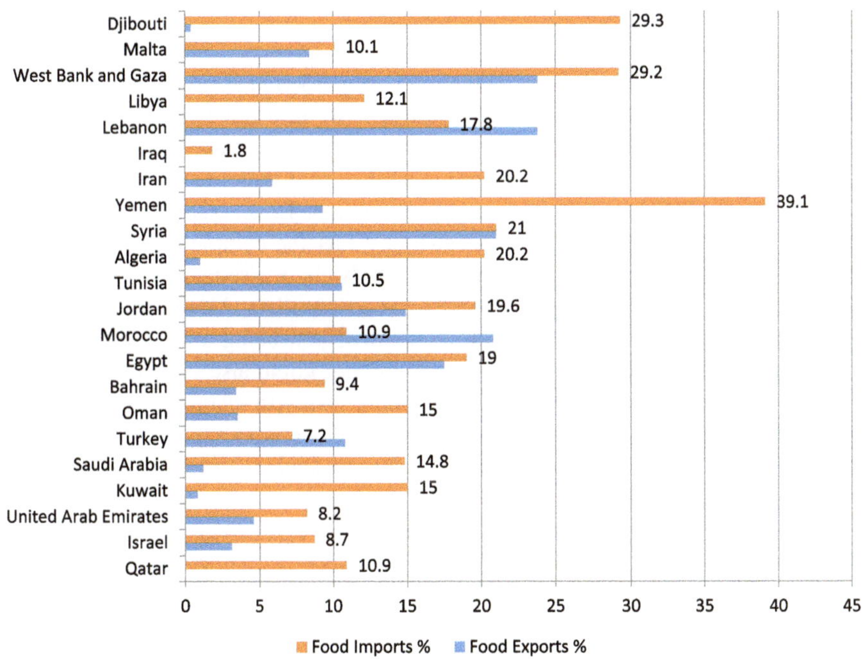

Fig. 11.6 Percentage of food imports and exports in 2019* (*As percentage of total merchandise imports and exports, respectively. Source: WDI (2020); http://wdi.worldbank.org/table/4.4#)

recessionary impact of the crisis, which is expected to lead to foreign demand contraction for their exports that are relatively highly priced and income elastic.

Countries in the region, mainly the GCC, depending heavily on exporting hydrocarbons are likely to experience a decrease in price of their major export commodities (mainly oil exports), which will in turn have a negative impact if price of staple food commodities continued to increase (Erokhin and Gao, 2020). Woertz (2020) classified those countries as the most vulnerable to COVID-induced trade disruptions as they are poor in natural resources as well as agricultural potential. Although the GCC countries' scores within the FS utilization pillar, as measured by GFSI (Table 11.1 in Annex), are relatively high, such countries hold the largest per capita shares worldwide for people suffering from diabetes and obesity (Woertz, 2020). It is worth mentioning that the GFSI has a major shortcoming in that respect as it does not cover a number of malnutrition and health-related indices. Nevertheless, the food systems of the GCC countries have so far performed well during the COVID-19 crisis in terms of ensuring food availability. Their modern value chains are more resilient than the traditional and transitional ones that dominate in developing countries (Woertz, 2020). In other MENA countries where the government plays a vital role in the procurement, storage, and distribution of main food products like wheat, the impact of the crisis on FS has been contained (e.g., Egypt) (Larson et al., 2012).

Tourism

One of the indirect transmission channels affecting FS is the slowing down of tourism in MENA region. Tourism is the main activity in a large number of MENA countries whether in terms of contribution to GDP or as a main direct and indirect employer. For example, tourism constitutes a major contributor to GDP and employment in several MENA countries as Egypt, Morocco, Tunisia, and Lebanon (FAO, 2020; Ghoneim, 2020). The slowing down or rather the complete stop of tourism could have severe repercussions on the food supply chain associated with hotels and restaurants, which have negative impact on food suppliers. Moreover, the cutting down of salaries and employees in the tourism sector, being a large employment sector in MENA, and especially for casual workers, could also have serious repercussions on the demand for food, which again is likely to disrupt the food supply value chain. According to some estimates the losses of tourism can constitute two-thirds of the total losses incurred by MENA countries due to the pandemic (OECD, 2020; World Bank Group, 2020).

Availability might not be radically affected in the region today but with the continuation of the pandemic along with its drastic recessionary impacts and escalating disruptions in the global and local supply chains, the crisis could have considerable impacts on availability, accessibility, and harshly on utilization in the very near future. Such potential shifts would have a dual effect on FS. The first effect arises from its negative impact on FS by shifting from healthy to less healthy food products, hence exacerbating the FI in the region. The second effect arises from the negative impact on the food trade balance since MENA countries export the high elastic food products (e.g., fruits, vegetables) and import the lower elastic food product (e.g., wheat, and grains). Hence the ability of the governments to ensure FS is expected to be weakened. This is likely to affect the balance of payments of the majority of MENA countries whether those highly depending on exporting hydrocarbons, importing low elastic food staples, or exporting high elastic food products. Conflict countries and LDCs are more vulnerable than other countries, but so as well are countries heavily depending on oil and/or food exports. This implies that the majority of MENA countries are highly vulnerable countries to the crisis. Table 11.4 in Annex illustrates the main sources of risk that countries in the MENA region could be exposed to due to the pandemic (FAO, 2020).

Policy Responses in MENA Countries

Some of the measures taken to confront and mitigate the repercussions of COVID on the food system were reactive and some were proactive. Among the reactive mitigating measures undertaken were expanding the cash transfers and consumer subsidies. Such measures helped to smoothen the expected potential price and income fluctuations in general and for food in specific (World Bank Group, 2020). The proactive measures included increasing the storage capacity. Food crisis 2007/2008 and the foresighted measures taken at that time to mitigate its impact was a main savior for MENA region at least during the early phases of the pandemic. Several MENA governments have learnt to add to their capacity of storing main food staples since the 2007/2008 crisis. This resulted in increasing the share of the global stock of the main food staples in the region and significantly increasing the stock-to-use ratio (Larson et al., 2012). All this accompanied with the favorable weather conditions for their major cereals and grains has set MENA region in a relatively stable position despite the Maghreb countries being subject to drought and above average weather temperatures as well as conflict in Syria, Yemen, Iraq, and Libya (FAO, 2020). This had helped in releasing the stress on the availability pillar in the region.

Along the same orientation of putting self-sufficiency as a main target, a number of countries in the region adopted trade protectionist measures, as a reactive measure, aiming at enhancing food imports or restricting/banning some food exports (Table 11.5 in Annex). Yet, the severity of such measures, and especially the export bans, decreased over time. Some MENA countries have dedicated extra attention to food aspects amid COVID-19 crisis, among which are Egypt (see Box 11.1 for measures adopted in Egypt) and Lebanon (special banking arrangements for importing food). The GCC countries have managed the crisis well so far, not through a reactive response policy but rather a proactive one. They have focused mainly on managing the supply chains and international linkages in an effective manner. Following Woertz (2020) and FAO (2020), they have increased storage capacity, diversified their importing sources, relaxed import procedures (e.g., relaxing the condition of Arabic labels), and streamlined sanitation measures. As a result of such

Box 11.1 Measures Taken to Shield Domestic Food Markets in Egypt

Egypt has ample food supplies to help shield its domestic market from short-term global supply risks associated with the COVID-19 crisis. As of April 2020, Egypt has already imported substantial quantities to cover its needs for the year, particularly wheat. It has, over the years, doubled its modern grain silo capacity (from 1.5 million tonnes in 2014 to 3 million tonnes in 2019). This favorable supply environment is supported by several government measures taken to enhance food availability in the country for at least several months:

- The Government imported significant quantities of basic staples including wheat, yellow corn, soybeans, and legumes.
- Egypt's cultivated wheat area this season exceeded 3.4 million feddan (1.38 million ha), and production is expected to be at least 9 million tonnes, which is the same level as last year and the 5-year average. Winter crops currently being harvested include barley, fava beans, alfalfa, potatoes, onions, and other vegetables. Sugar factories currently receive beet from farmers that will increase domestic sugar supplies.
- The Central Bank of Egypt (CBE) has expanded its EGP 100 billion industry stimulus initiative to include SMEs in fish, poultry, and livestock companies.
- About 29,000 families of community school students and teachers received unconditional cash transfers in nine governorates (Assiut, Aswan, Luxor, Giza, Fayoum, Matrouh, Qena, Bani Suweif, and Minia) redeemable for nutritious food items at local retail shops.

Sources: FAO (2020); WFP (2020b).

mitigating measures as well as enhancing production of strategic crops, the impact of COVID-19 on FS has been largely contained to date (FAO, 2020). Focusing on specific cases associated with COVID-19 crisis, a number of MENA countries have opted for additional policies. For example, the Lebanese government adopted a special arrangement allowing traders to use pegged exchange rates when purchasing grains from the world market, as a means to buffer the domestic market from the impact of the COVID-19 crisis and the financial problems Lebanon has been passing by (FAO, 2020).

Resorting to intensifying the usage of digitalization within food supply systems is another type of response to the crisis that is not efficiently utilized till today in the region. Digitalization has proved to shield many sectors or even aid them to thrive during the crisis. COVID has created a good opportunity to exploit such a method that would contribute to establishing resilient and sustainable food systems. For example, the utilization of technology in disseminating technical information to farmers as a means of outreaching a large number in remote areas while providing a regular follow-up can have significant positive impact on productivity. To the knowledge of the author, only some first steps for such orientation exist in Egypt where several initiatives for developing software applications for farmers are in progress (IFPRI, 2020b). The measures adopted by MENA countries helped to relax the availability constraint, despite the fact that it has not been the main concern whether on the global level (as depicted in *Section One*) or on the regional level as explained above. However, measures to tackle the accessibility pillar have been general and not food focused when it comes to raising the affordability of the population (e.g., social safety net measures). As identified by UNSDG (2020), the people who were already exposed to nutrition and dietary deprivations before COVID-19 crisis remained the ones most vulnerable after the crisis. Finally, measures tackling the utilization pillar remained rather absent.

Empirical Assessment of the Relationship Between the Pandemic and Food Security

Anecdotal evidence addressing the impact of the pandemic on FS has been quite extensive, yet empirical studies based on primary or secondary data has been relatively limited given the recency of the issue and the bundle of uncertainties carried within. Empirical studies included those depending on field surveys through interviews with households (Amare et al., 2020; Elsahoryi et al., 2020) and those using secondary data whether on country (Erokhin and Gao, 2020) or household levels (Amare et al., 2020). Some studies followed robust econometric analysis (Erokhin and Gao, 2020; Elsahoryi et al., 2020; Amare et al., 2020), while some followed scenario-based exploratory approach (Umdale et al., 2020) and others diverted to simple logical descriptive analysis (Mouloudj et al., 2020). Moreover, a set of IFPRI and World Bank studies employed a SAM multiplier and CGE models (Raouf et al., 2020; El Kadhi et al., 2020; Breisinger et al., 2020, World Bank Group, 2020) to assess the economy-wide impacts of the crisis.

This section addresses the nature of the link between the pandemic and FS through understanding the intercountry variability by applying simple cross-sectional econometric analysis. In a trial to understand this relationship within the constraint of data limitations due to the novelty of the issue, two approaches were followed. The first tries to understand the link between various country-specific capacities in handling the pandemic on one side and its FS level on the other. The second relates to the severity of the disease itself and the FI situation in each country. In the two approaches, the MENA situation is compared to the rest of the world.

First Approach

The first approach aims at understanding the link between COVID country preparedness (rather than the intensity of COVID occurrence per se as in the majority of the reviewed studies) and FS levels in various countries. The composite global FS index (GFSI[3]) is taken as a proxy for FI. The lower its score, the higher is the status of FI. The index covers a set of developing and developed economies amounting to 113 countries. The value of the index scores in 2019 represents the dependent variable that the model seeks to understand and the significance of the variables that explain the intercountry variability. The index covers all MENA countries except for Iran, Djibouti, Palestine, Iraq, Libya, Lebanon, and Malta.

To measure the effect of the countries' variability regarding their degree of readiness for handling and dealing with the disease, the composite INFORM COVID risk index, published by the European Commission, is used. It is an adapted version of the epidemic risk index, which provides scores for countries according to the degree of risk arising from a taxonomy of indicators capturing demographic, economic, socioeconomic, and institutional rigidities like, among others, the level of awareness, population density, proportion of population at increased risk of severe COVID disease, economic vulnerability, and access to healthcare as well as governance. These types of rigidities are lumped into three sub-indices, namely, the hazard and exposure, vulnerability, and lack of coping capacity indices. The index covers such sources of strains that existed before the pandemic outbreak in each country and that could highly limit effective national response to the crisis resulting in health and humanitarian hazards. The lower the score the better the country in each of the three dimensions in handling the pandemic (Poljanšek et al., 2020).

Since the vulnerability sub-index includes FI within its measurements, it was excluded from the analysis to avoid potential endogeneity problems. The analysis thus includes the other two sub-indices, namely, hazard and exposure and lack of coping capacity.

Yemen, Djibouti, Libya, Syria, Palestine, and Lebanon had the highest risk scores[4] within the MENA region for COVID coping capacities reflecting the institutional (governance and corruption) as well as health-related infrastructure deficiencies.[5] Bahrain, Yemen, Djibouti, Palestine, Oman, and Qatar had the worst scores compared to other MENA countries in COVID hazard and exposure index, which covers demographic aspects as well as sanitation and hygiene (Fig. 11.7).[6] It

[3] This is a comprehensive FS index that has been published by Economic intelligence Unit since 2012, which is a country-level food-security measurement tool that addresses the issues of affordability, availability and utilization (quality and safety) in 113 developed and developing countries.

[4] High scores were assumed the ones exceeding 5 out of 10, 10 is the highest and thus the most risky.

[5] This index includes health system capacity specific to COVID-19, government effectiveness, corruption perception index, immunization coverage, health system capacity, per capita public and private expenditure on healthcare, and maternal mortality ratio.

[6] This index includes population density, urban population growth, population living in slums and household size in addition to availability of drinking water, sanitation, and hygiene (Poljanšek

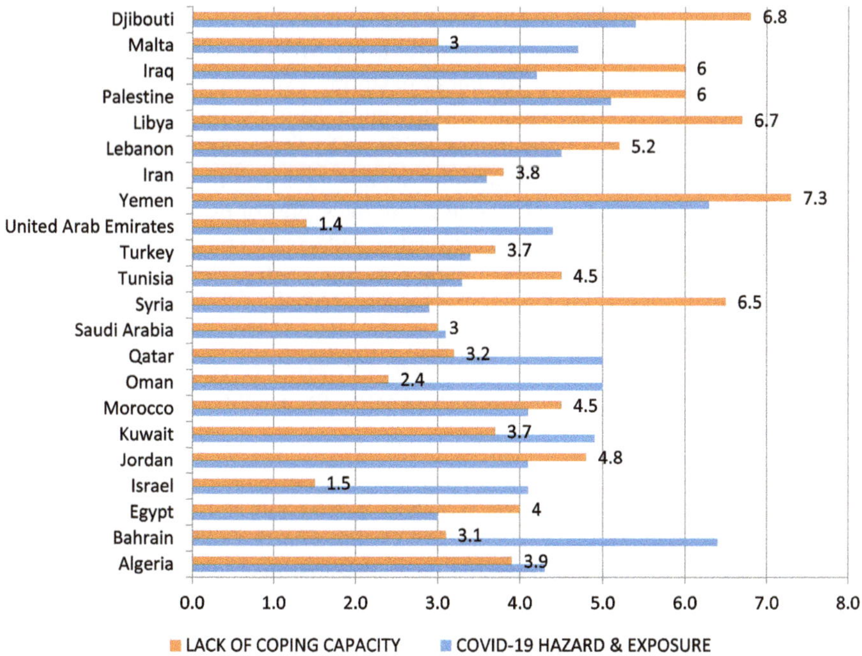

Fig. 11.7 COVID-19 Hazard and Exposure and Lack of Coping Capacity Risk Indices in MENA countries. (Source: Poljanšek et al. (2020))

is worth noting that Lebanon, Yemen, Palestine, and Djibouti have been classified as having "high" COVID risk regarding preparedness (grasping the three dimensions), whereas "Low" degree of risk characterized high-income countries in the region (Qatar, Saudi Arabia, UAE, Oman, and Israel) and all the rest lied in the "medium" range (Poljanšek et al., 2020).

To incorporate the effect of country variability within the various pillars of FS, the estimated equation includes two measures for food affordability and two measures to control for each of the accessibility and availability pillars. The ratio of food imports to total exports was included to reflect the affordability dimension. This variable illustrates a country's ability to purchase food from international markets using its export revenues, while also taking into account food availability and food accessibility on the international market (Breisinger et al., 2012). Food inflation rates were used to gauge for the accessibility dimension. GDP per capita was used to encompass a wider measure for the affordability dimension whereas the average dietary energy supply adequacy was included as a proxy to reflect food availability (Table 11.2).

The data set for this model includes all countries simultaneously covered by the EIU (2020) GFSI and the INFORM COVID index amounting to 112 countries. The

et al., 2020).

regression equation includes two dummy interactive terms for examining the significance of the effect if the country belongs to the MENA region along with their corresponding level of COVID readiness regarding exposure and lack of coping capacity risk measures. This would help to comprehend the relative impact of the variations in those two indices on the variability in FS levels in the region compared to rest of the world. Values for all variables are taken for the year 2019, which is the latest year available and is suitable to use with the scores of the two sub-indices of the INFORM COVID risk index as the latter reflect the country-specific demographic and institutional rigidities initially prevailing before the occurrence of the pandemic. Eq. (11.1) takes the following form:

$$\begin{aligned} GFSI_i = {} & \alpha_i + \beta_1 * HAZ_i + \beta_2 * COP_i + \beta_3 * GDPC_i + \beta_4 * DES_i \\ & + \beta_5 * FDINF_i + \beta_6 * RFDIMP_i + \beta_7 \left(MENA_i * HAZ_i \right) \\ & + \beta_8 \left(MENA_i * COP_i \right) \end{aligned} \tag{11.1}$$

Results: A two-stage least squares was used to overcome endogeneity problems.[7] Results showed that both coefficients of COVID-related risk indices had highly significant estimates with negative signs revealing that the higher the country is at risk of having weak demographic, health-related, and institutional factors in copying with the pandemic, the lower the score it gets for FS. This implies that the various existing structural rigidities in various countries that could limit their ability to handle the pandemic are negatively related to their FS level. As Table 11.3 shows, the coefficient of the lack of coping capacity in confronting the pandemic had a higher magnitude compared to that of hazard and exposure. Looking at the details of the measures upon which each sub-index is built, institutional aspects like weak governance and high corruption, as well as health-related infrastructural deficiencies that limit the ability to cope with the pandemic, are all related to higher levels of FI with a higher degree compared to demographic variables like population density, urban population growth, or population living in slums. This adds up more complexity to the transmission mechanisms being not only limited to supply and demand pathways (Table 11.3).

The significant coefficients for food inflation and GDP per capita (as measures for accessibility and affordability respectively) as well as the food availability proxy and their corresponding signs (Table 11.3) are in line with the conceptual framework discussed earlier. Coefficients of the MENA dummy interactive terms showed significant estimates for both indices with much worse results regarding the lack of coping capacities associated with lower levels of FS compared to the rest of the world. This puts the region at a disadvantaged situation regarding its coping capacity limitations as represented by the weak governance, corruption, and fragile health systems compared to the rest of the world. The coefficient concerning the exposure

[7] Instruments were tested and results rejected the null hypothesis of having weak instruments. Also the model was tested for over-identification (using Sargan and Bassman tests) and the p values led to not to reject the null hypothesis of having a valid set of instruments and that the model is correctly specified.

Table 11.3 Results of regression estimation (11.1), FS and COVID risk exposure and lack of coping capacities

Variables	Definition	Estimates
GDPC	GDP per capita at constant prices (=2015)	0.0001161** (0.003)
HAZ	COVID-19 Hazard and Exposure Index	−2.524814*** (0.000)
COP	Lack of coping capacity	−2.869009*** (0.000)
DES	Average dietary energy supply adequacy (percent) (3-year average) OR Average value of food production (constant 2004–2006 I$/cap) (3-year average)	0.0854127 *** (0.000)
FDINF	Food inflation	−0.3145633*** (0.000)
RFDIMP	Ratio of food imports to total exports	1.240247 (0.629)
MENAHAZ	Dummy variable that takes a value of 1 if the country belongs to the MENA region and 0 otherwise along with the hazard and exposure index	2.973736*** (0.000)
MENACOP	Dummy variable that takes a value of 1 if the country belongs to the MENA region and 0 otherwise along with the lack of coping capacity index	−3.369261*** (0.000)

***p 〈 0.01, **p 〈 0.05, *p 〈 0.1

and hazard index showed a counterintuitive positive link with FS, though with a negligible magnitude. Though it is difficult to interpret such result, the negligible effect implies highly weak relationship.

Second Approach

The second approach aims at investigating the link between the intensity of the pandemic per se after its outbreak in 2020 and FI through cross-sectional analysis. In line with a number of studies, the severity of the pandemic is measured in terms of the number of confirmed cases in each country (Adame et al., 2025, Erokhin and Gao, 2020). While there is no agreement in the literature regarding the suitable measure for FI, the study adopted the same definition applied by Erokhin and Gao (2020) where they expressed it by the number of people with poor or borderline food consumption as it was the only indicator reflecting FI amid the pandemic on a country level (Table 11.4).[8]

[8] This measure is defined as "People with insufficient food consumption refer to those with poor or borderline food consumption, according to the Food Consumption Score"; "Poor food consump-

Table 11.4 Definition of the new variables in Eq. (11.2) and their data sources

Variable	Definition	Data source
FI (dependent variable)	FI as measured by the number of people with insufficient food consumption per million in each country	World Hunger Map
COVNUM	COVID number of confirmed cases in each country	World Hunger Map

Data concerning both measures were collected from the World Hunger Map,[9] published by the WFP, as per the latest figures by the end of January 2021. The map registers the values of those variables, among others, on daily basis covering 93 developing economies that experience food consumption insufficiency. This infers that it does not cover the high-income economies in the MENA region, which includes GCC countries as well as Israel.

$$FI_i = \alpha_i + \beta_1 * COVNUM_i + \beta_2 * GDPC_i + \beta_3 * DES_i + \beta_4 * FDINF_i + \beta_5 * RFDIMP_i + \beta_6 \left(MENA_i * COVNUM_i \right) \tag{11.2}$$

Results: Applying ordinary least squares (OLS) on this cross-sectional data showed that the intensity of the pandemic had a significant positive coefficient, which reveals that countries experiencing high numbers of confirmed COVID cases are also having high numbers of people with poor or borderline consumption (Table 11.5). This is in line with the arguments raised in *Section Two* and all expectations of international organizations where the harsher the intensity of the disease, the higher the probability the country would fall in a FI status. Though the data sets are different where here it is only limited to developing countries while that used in Eq. (11.1) was with a wider scope covering a more non-homogeneous set of countries, yet comparing the two COVID-related impacts revealed that the degree of link between FI on one side and the demographic, institutional, and health-related deficiencies, on the other, exceeded by far that of the intensity of the disease per se (Table 11.5).

As Table 11.5 reveals, GDP per capita at constant prices has a negative significant coefficient estimate emphasizing the same direction of linkage with FS in Eq. (11.1), where the higher the GDP the less the FI level the country suffers from, since higher GDP per capita implies not only better affordability but also expected better diet and nutrition aspects. Note that the magnitude of the coefficient is much higher than that in Eq. (11.1) denoting the importance of GDP per capita in explaining the intercountry differences in FS levels within the set of developing countries suffering from relatively high level of FI. Unlike the results reached by Erokhin and Gao

tion: Typically refers to households that are not consuming staples and vegetables every day and never or very seldom consume protein-rich food such as meat and dairy. Borderline food consumption: Typically refers to households that are consuming staples and vegetables every day, accompanied by oil and pulses a few times a week." (https://hungermap.wfp.org/)

[9] https://hungermap.wfp.org/

Table 11.5 Results of regression estimation (11.2), FI, and COVID intensity (confirmed cases)

Variables	Estimates	
COVID intensity (confirmed cases)	0.000022*** (0.000)	Number of observations = 84 F(6, 77) = 52.69 Prob > F = 0.0000 R-squared = 0.8041 Adj R-squared = 0.7889 Mean VIF | 1.24 Breusch-Pagan/Cook-Weisberg test for heteroskedasticity Ho: Constant variance F(1, 82) = 0.30 Prob > F = 0.5859
GDP per capita at constant prices (=2015)	−0.0026219** (0.005)	
Average dietary energy supply adequacy (percent) (3-year average)	0.0108127 (0.906)	
Food inflation	−0.0033838 (0.802)	
Ratio of food imports to total exports	−8.341193 (0.305)	
MENA COVID intensity (confirmed cases)	−0.0000118** (0.042)	

***p ⟨ 0.01, **p ⟨ 0.05, *p ⟨ 0.1

(2020), food inflation reflecting accessibility dimension is insignificant, which can be explained by the usage of 1 year cross-sectional data in our case where price fluctuations are smoothened whereas Erokhin and Gao (2020) applied monthly data. However, they still reached some counterintuitive results (negative relationship) regarding the link between inflation and FI, which they explained by governments' intrusion in many developing countries in setting food prices (Erokhin and Gao, 2020). The ratio of food imports to total exports variable remained insignificant as in the first equation estimation. Like the first approach, the MENA interactive coefficient is significant yet weak (P value less than 0.1). Having a negative value with a lower magnitude than that of the number of confirmed COVID cases, reflects still a positive relation with FI in the MENA region (the higher the number of cases, the higher the degree of FI), yet with a slightly lower magnitude compared to the rest of the world (Table 11.5). This implies that the COVID-19 intensity affects the FI in MENA in negative terms when using the number of people with insufficient food consumption as an indicator yet with a degree less than other developing countries covered by the dataset. Although the study did not follow income country classification as Erokhin and Gao (2020) did, this result does not contradict theirs where they found FI in some MENA countries weak (Turkey) or even negatively linked to the number of COVID cases (Yemen, Iran, Jordan, and Libya). They found weaker link for low-income countries in general (Erokhin and Gao, 2020). Moreover, the weaker link between the number of COVID cases and the number of people suffering from insufficient food consumption for MENA might be in contrast to intuitive expectations that many countries within the region suffer relatively more from the modest accessibility and utilization pillars of their FS systems. Yet, when taking into account that the data does not include developed countries, the interpretation is clear where MENA countries (excluding high-income countries) is in a relatively better position when compared to other countries included in the set, which mainly includes Sub-Saharan and Latin American countries.

Limitations

In evaluating the study results, several limitations need to be taken into account. First, the cross-sectional nature of the study does not allow for causality to be inferred, nor was it feasible to assess changes over time, which was highly restricted by data insufficiency due to the novelty of the topic. Second, classifying countries according to their income level would have enriched the analysis, yet again this was restricted by data availability and would have resulted in lowering the number of observations. Third, estimations should not be taken as precise estimates in magnitude being based on cross-sectional data that ignores time dimension. Fourth, the World Hunger Map data, upon which the second equation is based, covered only those countries already experiencing high prevalence of undernourishment, which makes the analysis more restricted. Finally, the analysis treated MENA countries as one entity, though, as previously discussed, impact of the pandemic on FS has very country-specific effects within the region that should be accounted for, an area needed for further research.

Conclusion and Policy Implications

The study has shown that COVID-19 crisis when observed as a food crisis is significantly different from the food crisis in 2007/2008. Apart from the severity, the scope, the vagueness of its progress, time plays a pivotal role in the assessment of its repercussions. During the course of writing this chapter, information has changed and expanded on a daily basis bringing up many challenges to encompass the various dimensions of the crisis. More importantly, this draws attention to the significance of timely proactive solutions and responses to contain the divergent consequences of the crisis.

MENA countries are facing the COVID-pandemic with an already alarming FI situation that has been deeply rooted in a number of countries within the region. This deprives many countries from having full power to buffer the transmission of a multitude of adverse effects of the crisis on all FS pillars. COVID has mainly affected the affordability and utilization pillars of FS, and almost had negligible effects on the availability pillar, at least in the interim. In this regard, the credit goes partially to the food crisis in 2007/08, which triggered the increase in staples storage in most of the countries as a strategy that shielded many from a good fraction of the negative impacts of COVID on FS. A considerable portion of challenges facing MENA countries lies in the measures adopted to respond to the crisis where governments all over the world as well as in MENA region have reacted, so far, in a general and traditional manner. Their interventions have mainly focused on relaxing the least problematic constraint, availability, and left aside tackling the accessibility and utilization pillars. Depending on reserves and stocks, while it was a major savior for many countries, is not a sustainable policy. With the unforeseen duration of the crisis and its drastic recessionary impacts that are getting gloomier on daily basis, the availability pillar would be also hurt and countries should be ready for that scenario.

Wrongfully putting food sufficiency rather than FS as a main goal, trade restrictions came as a first resort to a number of countries just as they did as a panic reaction during the crisis in 2007/2008, though with a much lesser degree. A number of countries have imposed export restrictions or even bans on a list of food items and some have relaxed imports' restrictions but both served the availability pillar. Resorting to such trade restrictions, although proved that it injures more than cures throughout the experience in the 2007/08 food crisis, might only offer a very transitory solution.

MENA is not a homogeneous entity, the degree of exposure to food-related COVID risks differ considerably across countries in the region. GCC countries face a relative minimum degree of risk where the countries secured their food supply through ensuring enormous reserves, facilitating imports by giving up restrictive import requirements and minimizing the risk by diversifying providers. Moreover, their relatively limited agricultural production and low agricultural exports lowered the probability of being subjected to COVID-19-associated supply risks to domestic production. Further, their edge lies in having modernized well-established capital intensive value chains that significantly helped in mitigating the crisis impact. Realizing this fact, they responded by intensifying such chains and their international linkages. Although they have very strong affordability pillar, they still need to address a descending utilization pillar having one of the highest diabetes and obesity worldwide with no sign of improvement. Having the majority of the population, in some GCC countries, as migrant workers, "the invisible martyrs of the food system",[10] suffering from variable degrees of social and legal discrimination is another major jeopardy. In many cases they are not covered by social security schemes in addition to enduring restricted labor rights. All such factors need extreme scrutiny and reform in order to empower the ability of a considerable segment of the population in general and in the face of the current crisis in particular. Other countries in the region face a higher degree of risk due to the relatively high rates of food import dependency as well as the high food expenditure shares, which make them more prone to shocks on the demand side. This includes Egypt, Algeria, Yemen, Iraq, Jordan, and Palestine. From this perspective, the risk is a bit minimized for Morocco and Tunisia having relatively lower food import dependency. Other factors like conflict in Syria and Yemen, financial crises and political unrest in Lebanon, and instability in Iraq magnify the risk. Institutional factors like corruption, deficient and weak demographic aspects, and absence of efficient social security systems in many cases undermine the ability of governments in providing the needed protection from the consequences of COVID-19 especially for the most vulnerable, marginalized, and poor strata.

[10] Zurayk (2020)

MENA countries should make use of this crisis to tackle the structural imbalances affecting accessibility and utilization pillars of FS. Woertz (2020) and other scholars identified that COVID has brought the time for MENA countries in general and GCC countries in specific to shift their major focus from food availability to food accessibility and utilization. The former efforts in MENA countries have left those two dimensions of FS lagging behind. Measures to tackle the affordability and accessibility pillars have been general and not food focused. In this respect, the enhancement of social safety nets as a measure undertaken by all governments might have positive spillover effects on the accessibility pillar, but it remained short on targeting FS per se. Such enhancement should not include civil society charity initiatives as a main resort, if the aim is to at least maintain the level of FI. This channel will soon dry up if the pessimistic expectations of expanding economic downturn came into effect. Among the policy recommendations that can be implemented in this respect is nudging their population toward a healthy diet and nutrition prescriptions. This might be done through lowering subsidies directed to unhealthy food items and/or providing healthy relatively cheap substitutes. Given the fact that the pandemic brought the right time for consumers to value the importance of consuming healthy food, reforming social safety nets should be tilted to focus on enhancing such kind of awareness. Moreover, MENA countries should devote extra efforts to fix their local supply chains and enhance their integration in global value chains when it comes to food products and inputs, while developing the accompanying risk management systems and overcoming logistical problems (FAO CFS HLPE, 2020; Mouloudj et al., 2020).

As identified by UNSDG (2020) the people who were already exposed to nutrition and dietary insufficiency before COVID-19 crisis remained the ones most vulnerable after the crisis. Perhaps the general impact of the crisis and dealing with it as a health and economic crisis has left FS at the back seat. This is an issue that needs to be prudently addressed by MENA governments to avoid its potential negative consequences. Anecdotal evidence (e.g., Woertz, 2020; Ghoneim 2015) for a large number of Arab countries indicates that MENA countries suffer from inefficiency in the domestic food markets, arising from anticompetitive behavior and fragmented supply chains. The measures undertaken by governments in MENA to contain COVID-19 effects did not focus on reforming such issues. COVID has brought the urgency to pursue radical reforms and revisit many directly and indirectly related structural and institutional long rooted rigidities across the region.

In order to understand the link between FI and the pandemic, the study employed two types of datasets and correspondingly two equations were estimated using two different FS indicators and two indicators to proxy the effect of the pandemic. The empirical assessment showed that COVID, whether expressed by its intensity, number of confirmed cases, or by the degree of country preparedness to handle the disease, is significantly linked the country variations regarding their FS levels. The higher the number of confirmed cases and the weaker the degree of country preparedness, the higher is the level of country FI level. However, as Zurayk (2020) constructively articulated, COVID is to a great extent innocent from the charges thrown on it of being the exclusive cause of many of the adverse impacts on FS. The pandemic has been rather the magnifying loop that made the inadequacy, caveats, and fragility of the food systems worldwide and in MENA countries, specifically, clear and more tangible. Along the same lines, the empirical assessment in the study showed that as much as FI variability across all countries is explained by the variability of the intensity of the disease per se, it has drawn vivid attention to the importance of the efficient institutional and demographic prerequisites needed to handle the disease in this respect. More precisely it revealed that the higher the country is at risk of having weak demographic, health-related and institutional factors in copying with the pandemic, the lower the score it gets for FS. Institutional aspects like weak governance and high corruption, as well as health-related infrastructural deficiencies that limit the ability to cope with the pandemic, came in the first place in explaining country variability for having higher levels of FI. Demographic variables like population density, urban population growth, or population living in slums came in the second place. Regression results have also put the MENA region at a disadvantaged situation, compared to rest of the world, regarding its coping capacity limitations as represented by the weak governance, corruption, and fragile health systems in explaining the country variability in FS levels. So again it is not the pandemic per se, but rather the structural, demographic, economic, and institutional deficiencies that attention should be devoted to in order to have a more sound level of FS, even if COVID did not exist.

The importance of the affordability pillar was also illustrated in the empirical part of the study where GDP per capita as a measure of affordability in both models has proved to significantly explain country variability regarding FS levels. The significant coefficients for food inflation as well as the food availability proxy and their corresponding signs within the broader set of countries, encompassing developing as well as developed, are in line with the conceptual framework showing their importance in understanding country FS variability. However, both pillars were insignificant in explaining the FS variability across developing countries already suffering from poor or borderline levels of insufficient food consumption. Within the latter set of countries, FI in the MENA region was also negatively related to the intensity of the disease (number of confirmed cases) yet with the lower degree compared to Sub-Saharan and Latin American countries covered by the dataset. This is in line with results of other studies like Erokhin and Gao (2020), where they found FI in some MENA countries weakly (Turkey) or even negatively linked to the number of COVID cases (Yemen, Iran, Jordan, and Libya).

Preferential trade agreements signed by MENA countries need to address FS. So far the existing trade agreements signed by MENA countries among themselves (e.g., PAFTA, Agadir) or with their major trading partners (as EU Association Agreements) have suffered from protectionist attitude when it comes to agriculture (World Bank, 2020). Revisiting the agreements needs to be considered with the aim of targeting FS and addressing lessening the protectionist attitude and engagement in secure and safe value chains.

A further key consideration is the availability and transparency of data. In order to hold a better understanding of the repercussions of the pandemic and thus offer rigorous analysis and constructive policy implications, more data transparency is needed. Within the region, a number of countries stopped declaring COVID testing data and fatality rates. For example, as World Bank (2021) noted, Algeria stopped reporting such figures when its positive rates reached extreme limits, while Syria did not report testing data at all (World Bank, 2021). In this respect the GCC countries are the best performers. This alone indicates that MENA faces a transparency challenge. Transparent dissemination of information in general regarding testing, number of confirmed COVID cases, food prices, and food situation will strengthen government management over the food market during such abnormal circumstances and lessen the degree of panic that usually results in unfavorable outcomes.

From another angle, COVID-19 has shifted the whole world to a new norm where digitalization became at the core element of personal and business livelihoods. FS still lags behind in this regard. Except for some early signals in Egypt, agriculture is not monitored by any digital application in the majority of MENA. Introducing information technology-based applications in farming, for example, whether in terms of enhancing awareness of small-scale farmers or for collecting and analyzing data is among the reforms that should be seriously considered in the region. The effects of such orientation are highly expected to be translated into improved productivity as well as intensified supply chains and better logistics. The use of information technology has opened a new set of businesses in the food industry through the use of applications for food order and delivery that could potentially and partially offset the job losses due to the lockdowns and confinements. In addition, COVID entailed a positive element with the improvement of many aspects of food safety, including delivery and dining behavior, provided that this would prevail after the crisis. The hope thus lies in the pandemic itself to trigger the long-lasting neglected reforms within food systems and across the indirectly related channels in the same manner by which 2007/2008 food crisis unintentionally cushioned major repercussions of COVID today.

References

Adame, B. J., Corman, S. R., Von Feldt, P. A., Meneses, C. M., O'Rourke, H. P., Tahir, A. (2025). Reactance as a cause of COVID-19 vaccination hesitancy. Vaccine (57): 127209. https://doi.org/10.1016/j.vaccine.2025.127209.

Amare, Mulubrhan, Kibrom A. Abay, Luca Tiberti and Jordan Chamberlin (2020), Impacts of COVID-19 on Food Security: Panel Data Evidence from Nigeria, August 2020; PEP Partnership for Economic Policy., https://www.ifpri.org/publication/impacts-covid-19-food-security-panel-data-evidence-nigeria

Barrett, C. (2010). Measuring food security. *Science, 327*, 825. https://doi.org/10.1126/science.1182768. https://www.researchgate.net/publication/41424901_Measuring_Food_Security

Breisinger, Clemens, Jean-Francois Maystadt, Jean-Francois Trinh Tan (2012), *Does food security matter for transition in Arab countries?*., IFPRI Discussion Paper 01196, Development Strategy and Governance Division, July 2012; https://www.researchgate.net/profile/Clemens_

Breisinger/publication/254416880_Does_Food_Security_Matter_for_Transition_in_Arab_Countries/links/55b8bf8008ae092e965a3d78.pdf

Breisinger, C., Ecker, O., Maystadt, J.-F., Tan, J.-F. T., Al-Riffai, P., Bouzar, K., Abdelkarim, S. M. A., & Abdelgadir, M. (2014). *How to build resilience to conflict, the role of food security*. Food Policy Report. https://www.ifpri.org/publication/how-build-resilience-conflict-role-food-security

Breisinger, C., Raouf, M., Wiebelt, M., Kamaly, A., & Karara, M. (2020). *Impact of COVID-19 on the Egyptian economy: Economic sectors, jobs, and households*. IFPRI. https://www.ifpri.org/publication/impact-covid-19-egyptian-economy-economic-sectors-jobs-and-households

Brinkman, H.-J., de Pee, S., Sanogo, I., Subran, L., & Bloem, M. W. (2010). High food prices and the global financial crisis have reduced access to nutritious food and worsened nutritional status and health. *Journal of Nutrition*. https://doi.org/10.3945/jn.109.110767

Mandour Dina (2017), Food Insecurity and Political Unrest in Egypt: The Inter-Linkages, Paper presented at the conference Crisis and Conflict in the Agrarian World: An Evolving Dialectic, March 1–3, 2017, Sciences Po, .

Economist Intelligence Unit EIU. (2020). *Global Food Security Index*. http://foodsecurityindex.eiu.com/

El Kadhi, Z., Elsabbagh, D., Frija, A., Lakoud, T., Wiebelt, M., & Breisinger, C. (2020). *The impact of COVID-19 on Tunisia's economy, agri-food system, and households*. IFPRI. https://www.ifpri.org/publication/impact-covid-19-tunisias-economy-agri-food-system-and-households

Elsahoryi, N., Al-Sayyed, H., Odeh, M., McGrattan, A., & Hammad, F. .(2020). Effect of Covid-19 on food security: A cross-sectional survey, 2020 European Society for Clinical Nutrition and Metabolism. Published by Elsevier; https://doi.org/10.1016/j.clnesp.2020.09.026; https://www.sciencedirect.com/science/article/abs/pii/S2405457720302138

Erokhin, V., & Gao, T. (2020). Impacts of COVID-19 on trade and economic aspects of food security: Evidence from 45 developing countries. *International Journal of Environmental Research and Public Health, 2020*(17), 5775. https://doi.org/10.3390/ijerph17165775. https://www.mdpi.com/1660-4601/17/16/5775

FAO. (2020). COVID-19 and the impact on food security in the Near East and North Africa: How to respond? *Cairo*. https://doi.org/10.4060/ca8430en

FAO Committee on World Food Security High Level Panel of Experts (FAO CFS HLPE). (2020). *Impacts of COVID-19 on food security and nutrition: Developing effective policy responses to address the hunger and malnutrition pandemic*. http://www.fao.org/3/cb1000en/cb1000en.pdf

FAO, IFAD, UNICEF, WFP, & WHO. (2020). *The state of food security and nutrition in the world 2020. Transforming food systems for affordable healthy diets*. FAO. https://doi.org/10.4060/ca9692en

FAOSTAT. (2020a). Suite of food security indicators. http://www.fao.org/faostat/en/#data/FS

FAOSTAT (2020b), Data on Islamic Republic of Iran., http://faostat.fao.org/static/syb/syb_102.pdf

Ghoneim, A. F. (2015). The political economy of food price policy in Egypt. In P. Pinstrup-Anderson (Ed.), *Food price policy in an era of market instability*. Oxford University Press.

Ghoneim, A. F. (2020). *Exploring the potential impact of COVID-19 on trade in the Arab Region*. op-ed for Arab Development Portal. https://arabdevelopmentportal.com/blog/exploring-potential-impact-covid-19-trade-arab-region

Glauber, J., Laborde, D., Martin, W., & Vos, R. (2020). COVID-19: Trade restrictions are worst possible response to safeguard FS. *IFPRI Blog: Issue Post*. https://www.ifpri.org/publication/covid-19-trade-restrictions-are-worst-possible-response-safeguard-food-security

GRFC. (2020). Global report on food crises, joint analysis for better decisions.; https://docs.wfp.org/api/documents/WFP-0000114546/download/?_ga=2.70676240.1644043769.1612853053-1731074169.1572183068

IFPRI. (2016). Global Nutrition Report: From promise to impact ending malnutrition by 2030; http://www.ifpri.org/publication/global-nutrition-report-2016-promise-impact-ending-malnutrition-2030

IFPRI. (2020a). *Cash transfers as an effective tool for social protection and shock response in Egypt*. Symposium Policy Note 23 USAID and IFPRI. https://www.ifpri.org/publication/symposium-policy-note-3-cash-transfers-effective-tool-social-protection-and-shock

IFPRI. (2020b). *Digitalization of agricultural services and policy analysis in Egypt*. Symposium Policy Note 2, USAID and IFPRI. https://www.ifpri.org/publication/symposium-policy-note-2-digitalization-agricultural-services-and-policy-analysis-egypt

ITC. (2021). Market Access Map, COVID-19 temporary trade measures., https://www.macmap.org/Covid19

Karasapan, O. (2020). Striving for water and food security. *The Cairo Review of Global Affairs*. https://www.thecairoreview.com/essays/striving-for-water-and-food-security/

Kinsey, E. W., Kinsey, D., & Rundle, A. G. (2020). COVID-19 and food insecurity: An uneven patchwork of responses. *Journal of Urban Health, 97*, 332–335. https://doi.org/10.1007/s11524-020-00455-5. Published online June 2020. The New York Academy of Medicine 2020.

Larson, Donald F., Julian Lampietti, Christophe Gouel, Carlo Cafiero, & John Roberts. (2012). *Food Security and Storage in the Middle East and North Africa, The World Bank Development Research Group Agriculture and Rural Development Team April 2012, Policy Research Working Paper 6031*; https://www.researchgate.net/publication/254072586_Food_Security_and_Storage_in_the_Middle_East_and_North_Africa

Ling Ma, N., Peng, W., Soon, C. F., Hassim, M. F. N., Misbah, S., Rahmat, Z., Yong, W. T. L., & Sonne, C. (2021). Covid-19 pandemic in the lens of food safety and security. *Environmental Research, 193*, 110405. https://www.sciencedirect.com/science/article/pii/S0013935120313025

Petetin Ludivine (2020), *The COVID-19 Crisis: An Opportunity to Integrate Food Democracy into Post-Pandemic Food Systems*, Downloaded from https://www.cambridge.org/core. IP address: 196.155.13.63, on 22 Jul 2020 at 17:07:06, subject to the Cambridge Core terms of use, available at https://www.cambridge.org/core/terms. https://doi.org/10.1017/err.2020.40

Mouloudj, Kamel, Ahmed Chemseddine Bouarar, and Hamid Fechit (2020), "The Impact of Covid-19 Pandemic on Food Security", Les Cahiers du Cread, Vol. 36 - n° 03–2020 159, https://www.researchgate.net/publication/343047126_The_Impact_Of_Covid-19_Pandemic_On_Food_Security

Napoli, M. (2011). *Towards a Food Insecurity Multidimensional Index (FIMI)*. Roma TRE Universita Tegli Studi. http://www.fao.org/fileadmin/templates/ERP/uni/FIMI.pdf

Nasrin, O., & Ahmadi, D. (2019). Kate Sinclair and Hugo Melgar-Quiñonez (2019), Food Security in selected Middle East and North Africa (MENA) countries: an inter-country comparison. *FS, 11*, 531–540. https://doi.org/10.1007/s12571-019-00935-w. https://ideas.repec.org/a/spr/ssefpa/v11y2019i3d10.1007_s12571-019-00935-w.html

OECD (2020), COVID-19 Crisis Response in MENA countries, https://www.oecd.org/coronavirus/policy-responses/covid-19-crisis-response-in-mena-countries-4b366396/

Oraikat, S. (2020). *Corona and Food Security (in Arabic)*, Series of Papers Institute of National Planning, Cairo. *May, 2020*. http://repository.inp.edu.eg/bitstream/handle/123456789/4850/6.pdf?sequence=1&isAllowed=y

Pinstrup-Anderson, P. (2015). An Overview. In P. Pinstrup-Anderson (Ed.), *Food Price Policy in an Era of Market Instability*. Oxford University Press.

Poljanšek, K., Vernaccini, L., & Ferrer, M. (2020). *INFORM Covid-19 Risk Index*. European Commission. https://op.europa.eu/en/publication-detail/-/publication/be4524bf-a9f7-11ea-bb7a-01aa75ed71a1/language-en

Raouf, M., Elsabbagh, D., & Wiebelt, M. (2020). Impact of COVID-19 on the Jordanian Economy Economic sectors, Food Systems, and Households, Regional Program Policy Note 09. *November, 2020*. https://ebrary.ifpri.org/utils/getfile/collection/p15738coll2/id/134132/filename/134343.pdf

United Nations UN (2020a), Policy Brief: The Impact of COVID-19 on the Arab Region, An Opportunity to Build Back Better, July 2020a; https://www.un.org/sites/un2.un.org/files/sg_policy_brief_covid-19_and_arab_states_english_version_july_2020.pdf

United Nations UN. (2020b). *The Impact of COVID-19 on Food Security and Nutrition*, Policy Brief, https://Namibia.un.org/sites/default/files/2020-06/sg_policy_brief_on_covid_impact_on_food_security.pdf. *June, 2020b*.

UNSDG. (2020). The Impact of COVID-19 on the Arab Region An Opportunity to Build Back Better, Policy Brief. *JULY, 2020*. https://unsdg.un.org/resources/policy-brief-impact-covid-19-arab-region-opportunity-build-back-better

WDI (2020), World Development Indicators; https://databank.worldbank.org/source/world-development-indicators;

Woertz, E. (2020). Wither the Self-sufficiency Illusion? Food Security in Arab Gulf States and the impact of COVID-19. *FS, 12*, 757–760. https://link.springer.com/article/10.1007/s12571-020-01081-4

World Bank (2021), MENA Crisis Tracker, 2/2/2021, http://documents1.worldbank.org/curated/en/280131589922657376/pdf/MENA-Crisis-Tracker-February-2-2021.pdf

World Bank Group. (2020). Trading together: Reviving Middle East and North Africa Regional Integration in the Post-Covid Era, World Bank Middle East and North Africa Region. *MENA Economic Update October, 2020*. https://openknowledge.worldbank.org/handle/10986/34516

World Food Program WFP. (2020). *Economic and Food Security Implications of the COVID-19 Outbreak: The cost of the attempt to contain a highly contagious disease.*, Vulnerability analysis and Mapping Service VAM; https://fscluster.org/sites/default/files/documents/wfp-economic_and_food_security_implications_of_the_covid-19_outbreak.pdf

World Food Program WFP. (2020b). *WFP Egypt Country Brief*. https://reliefweb.int/report/egypt/wfp-egypt-country-brief-may-2020

World Food Program WFP. (2021), World Hunger Map, HungerMap Live; https://hungermap.wfp.org/

World Resources Institute. (2013). Water Stress by Country. *December, 2013*. https://www.wri.org/data/water-stress-country

Zurayk, R. (2020). Pandemic and Food Security: A View from the Global South. *Journal of Agriculture, Food Systems, and Community Development, 9*(3), 17–21. https://doi.org/10.5304/jafscd.2020.093.014. https://www.foodsystemsjournal.org/index.php/fsj/article/view/803

Chapter 12
Aligning Urban Farming with Green City Aspirations

John R. Groenewegen

Abstract Demands of a green city include many goods and services, such as: (1) fresh produce, (2) reduced food miles, (3) lower overall environmental footprint of food consumed in the cities, (4) ability to return organic waste generated in the city back to the soil, (5) biodiversity of the urban fringe lands, (6) access to open spaces in the urban fringe, (7) agri-tourism and other rural experiences, and (8) recreational activities in the urban fringe. While there are many opportunities for farmers to satisfy food-related demands of green city residents, there are also challenges, as it competes with other land uses, such as open space for recreation and land required for necessary infrastructure to support the metropolis. Peri-urban agriculture can include field production of fruits and vegetables, greenhouse production, and vertical farms. Greenbelts around a metropolis, which restrict certain types of land use and/or restrict changes in zoning from agriculture to other uses, are needed to support expansion of peri-urban farming.

Keywords Green city · Peri-urban agriculture

Introduction

Farm operations in the urban fringe provide food to the adjacent metropolis such as the Greater Toronto Hamilton Area (GTHA) in Ontario, Canada. Such farm operations that supply fresh, high-quality food into the metropolis can be characterized as peri-urban farming (or peri-urban agriculture). Food producing initiatives within the confines of the metropolis, such as community gardens, roof top-gardens, and vertical garden walls are part of urban farming and not considered here as part of peri-urban farming. This chapter focuses on food producing activities in the urban fringe, or peri-urban farming, as illustrated in Fig. 12.1.

J. R. Groenewegen (✉)
JRG Consulting Group, Guelph, ON, Canada
e-mail: john@jrgconsulting.com

A. A. Kubursi et al. (eds.), *Urban Food Security in a Crisis Prone World*,
https://doi.org/10.1007/978-3-031-89440-4_12

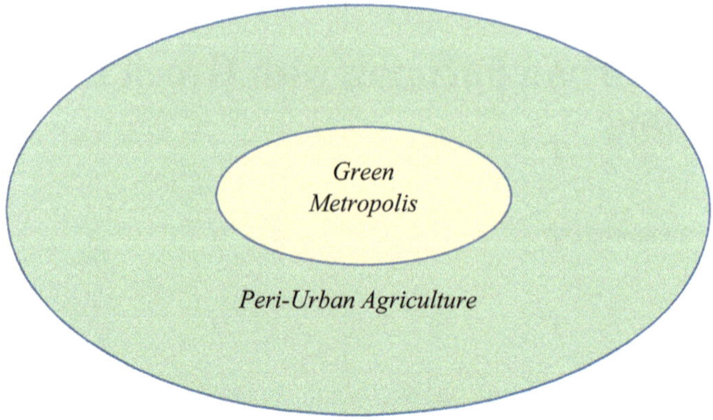

Fig. 12.1 The green city metropolis and peri-urban agriculture

A green city, including a green metropolis, can be described as a community of residents, workers, and visitors who strive together to balance ecological, economic, and social needs to ensure a clean, healthy, and safe environment for all members of society and for generations to come. An example of a green metropolis is the urban area of the Greater Toronto and Hamilton Area (GTHA) in southern Ontario. The dense urban area of the GTHA, with a population of more than 7 million, is shown in Fig. 12.2 as the purple shaded area along the north-western shore of Lake Ontario.

There is a symbiotic relationship between residents of a green metropolis and agriculture in the urban fringe. Green city dwellers want access to high quality, locally grown produce supplied by farm operations in the urban fringe and at the same time many farm businesses in the urban fringe invest in meeting dietary needs of the urban population.

The counties and regions (Durham, York, Toronto, Peel, Halton, Hamilton, and Niagara) that encompass the GTHA are referred to as the Golden Horseshoe (GH). In the GH, the 2016 Agricultural Census reported 196,465 hectares of cropland, with 17,750 hectares of fruit and vegetable crops. The labeled counties and regions in Fig. 12.2 are referred to as the Greater Golden Horseshoe (GGH), which had 14,350 hectares in fruit production and 20,250 hectares in vegetable production. There is room for expansion of fruit and vegetable production in this urban fringe region, given the 1.1 million hectares of cropland in the GGH. In 2016, there were 356 fruit farms and 713 vegetable farms in the GH and this expands to 894 fruit farms and 872 vegetable farms in the GGH.[1]

[1] Farm numbers are based on fruit or vegetable production accounting for the majority of farm sales. The actual number of fruit and vegetable farms can be greater than the numbers provided.

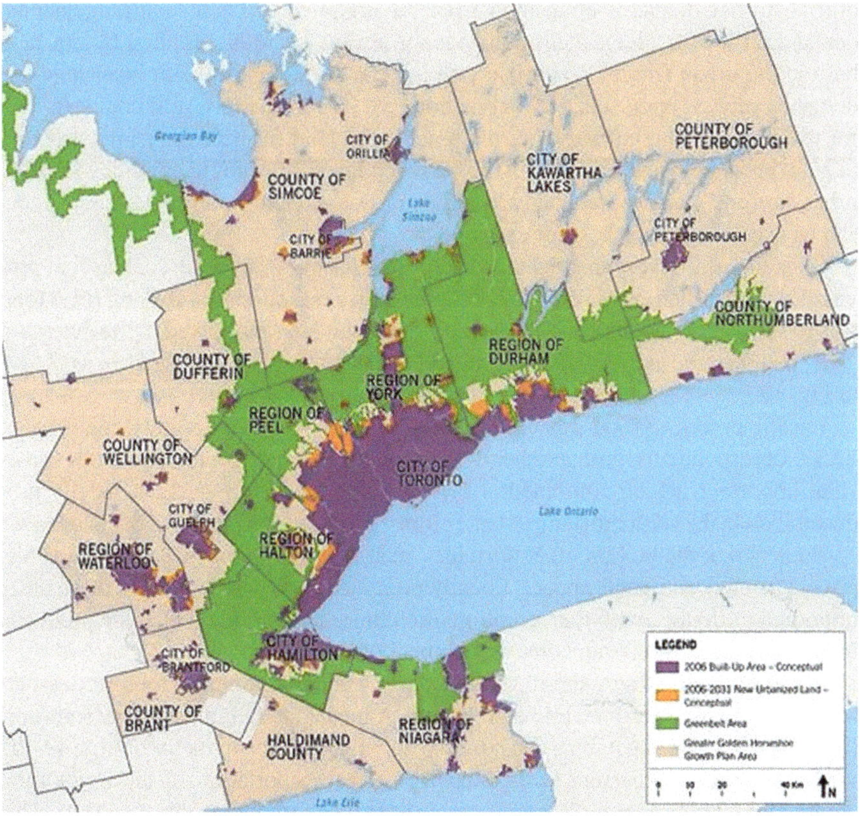

Fig. 12.2 The GTHA, the Golden Horseshoe (GH) and the Greater Golden Horseshoe (GHH) regions. (Source: Urban Toronto)

Aspirational Demands of the Green Metropolis on the Urban Fringe

Residents of a green metropolis have a demand for locally grown fresh high-quality foods that can be supplied from the urban fringe, which are typically fresh fruits and vegetables. The nature of demand that green cities place on the urban fringe extends beyond that for fresh produce and other foods. These green city demands on the urban fringe also include goods and services such as: (1) reduced food miles, (2) lower overall environmental footprint of food consumed in the cities, (3) ability to return organic waste generated in the city back to the soil, (4) biodiversity of the urban fringe lands, (5) access to open spaces in the urban fringe, (6) agri-tourism and other rural experiences, and (7) recreational activities in the urban fringe.

This multidimensional demand by the green city for goods and services in the urban fringe intersects with the ability of the land base surrounding the metropolis to supply these food requirements from farming operations and as well supply the other land use demands emanating from the urban core. These multidimensional demands from the green metropolis cannot always be fully supplied by the land base in the urban fringe. Peri-urban farming can be in conflict with other land use demands such as open space for recreation (e.g., walking trials, golf courses), land for residential and commercial development, and land for necessary infrastructure (e.g., expressways and airports). At times, such land use is not in conflict with peri-urban farming since some lands are fragile or not suitable for crop agriculture, but can be used as open spaces, for example.

Canadians have very distinct reasons for purchasing local food, such as that provided by peri-urban agriculture. In 2010, the major reason for purchasing local food was to support the local economy (at 53% of respondents) followed by freshness (at 29%) and then the associated lower overall environmental impact (at 7%), as shown in Fig. 12.3.

Locally produced food was noted in another survey to be extremely important to 42.5% of respondents, with another 34.5% indicated locally produced food is somewhat important, which combined is 77%. As illustrated in Fig. 12.4, locally produced food is the sixth most important priority guiding food purchase decision, after consumers tick the box on other attributes such as quality, safety, nutritional value, price, and environmental impact. Locally produced food can deliver on these other important attributes, with the exception often times being price. For some products, locally produced foods can come at a higher cost structure.

Peri-urban agriculture significantly reduces the environmental footprint associated with shipping produce into a green city. A recent study indicated that transportation of fruits and vegetables consumed within Ontario accounts for 6% (or 49,405 tons) of the CO_2 emissions associated with the transportation of all agricultural commodities grown within Ontario, as well as those imported into Ontario.[2] This

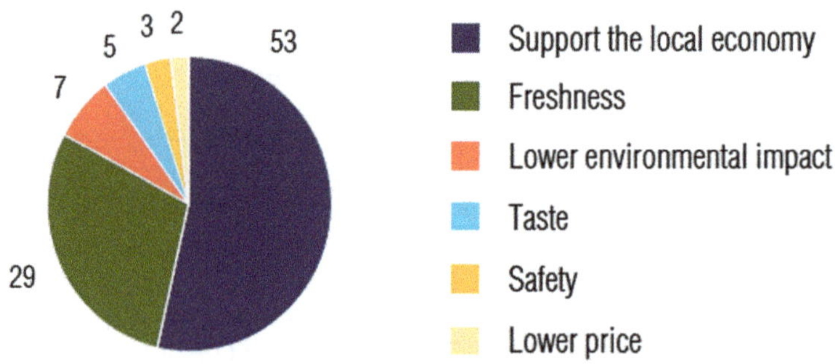

Fig. 12.3 Most important reasons for buying local food (percentage of respondents). (Source: Agriculture and Agri-Food Canada "*Consumer Perceptions of Food Safety and Quality: Wave 3 Tracking 2010*" Ottawa (June 2010), prepared by Ipsos-Reid Corporation.)

[2] "*DOLLARS & SENSE: Opportunities to Strengthen Southern Ontario's Food System*", prepared by Atif Kubursi, Harry Cummings & Associates, and Rod MacRae, PhD for the Greenbelt Foundation ISBN 978-1-927075-11-1 (January 2015).

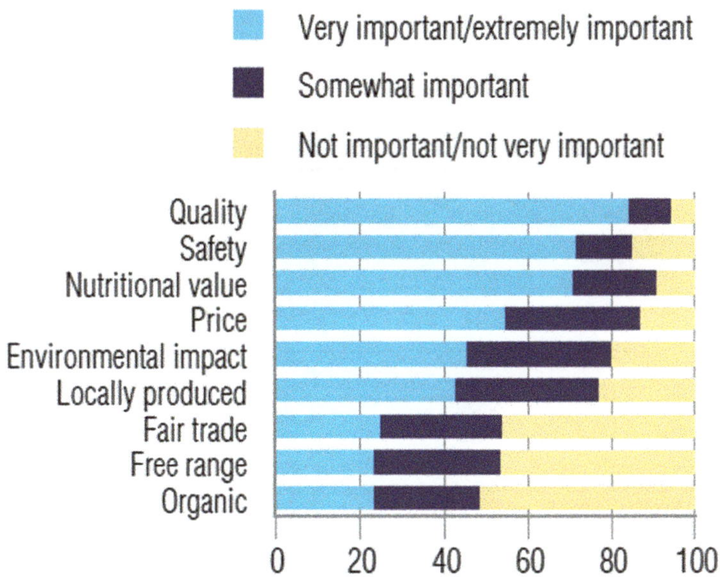

Fig. 12.4 Priorities when purchasing food (percentage of respondents). (Source: The Conference Board of Canada "*Cultivating Opportunities: Canada's Growing Appetite for Local Food*" prepared for the Conference Board of Canada's Centre for Food in Canada (CFIC) (2013))

same study indicated that a 10% reduction in fruit and vegetable imports would result in a 12,249 tons reduction in CO_2 emissions associated with the transportation of these imported commodities. The fewer road-miles associated with meeting green city requirements by peri-urban agriculture, the smaller the environmental footprint associated with transportation from the farm to the city.

Advantages to Entrepreneurs in Peri-Urban Agriculture

There are clear demands from the green city metropolis for locally grown fresh fruits and vegetables provided by farmers in the urban fringe. As a result, there are opportunities for entrepreneurial farm operations in the urban fringe to supply the green city requirements for fresh produce.

Using the GH metropolitan area, with an overall population of 7.3 million, peri-urban agriculture (within the GH) surrounding the metropolis has the production base to supply a significant portion of fresh fruit and vegetable requirements required by the metropolis. For example, strawberry production within the GH supplies 33% of consumption requirements for strawberries in a typical 3-month summer season (as shown in the first column in Table 12.1). In contrast, carrot production within the GH accounts for 391% of market requirements for fresh carrots over a 3-month period in the GH.

Table 12.1 Current ability of the fruit and vegetable production base, in the Golden Horseshoe (GH) and Greater Golden Horseshoe (GGH) regions, to supply residents of the GH and GHH with selected fresh fruit and vegetables

Crop	Current production in the GH as a percent of consumption for 7.3 million residents in the GH for a 3-month summer	Current production in the GGH as a percent of consumption for 9.1 million residents in the GGH for a 3-month summer	Current production in the GGH as a percent of annual fresh market requirements for 9.1 million residents in the GGH
Apples	175%	334%	84%
Peach	795%	647%	162%
Pears	80%	71%	18%
Strawberries	33%	63%	16%
Asparagus	6%	344%	86%
Beans (Green & Wax)	23%	242%	60%
Beets	276%	543%	136%
Broccoli	46%	55%	14%
Cabbage	94%	287%	72%
Carrots	391%	598%	150%
Cauliflower	87%	79%	20%
Celery	184%	148%	37%
Cucumber	32%	207%	52%
Dry Onion	166%	308%	77%
Green Peas	77%	363%	91%
Peppers	26%	201%	50%
Potatoes	68%	423%	106%
Pumpkins	117%	305%	76%
Radishes	50%	207%	52%
Rutabagas & Turnips	7%	96%	24%
Spinach	12%	13%	3%
Sweet Corn	225%	566%	142%
Tomatoes	73%	283%	71%

Source: Calculations based on county profiles prepared by the Ontario Ministry of Agriculture Food and Rural Affairs (OMAFRA) available at http://www.omafra.gov.on.ca/english/stats/county/index.html

Note: In-season fresh consumption requirements are based on the assumption of uniform consumption throughout the year, which will not hold for some seasonal crops such as peaches, strawberries, and sweet corn. For example, the 33% of a 3-month summer consumption requirement is likely much less once the seasonality of strawberry consumption is taken into account. The crops highlighted in green can be stored and are not as perishable as some more tender and perishable fruits and vegetables such as peas and peaches

Extending the urban fringe to include the GGH (see the map in Fig. 12.2 above), peri-urban agriculture in this region currently produces more than the needed for a 3-month seasonal requirement for most perishable vegetables. As illustrated in the middle column in Table 12.1, this includes asparagus, snap beans, celery, cucumbers, green peas, peppers, pumpkins, sweet corn, and tomatoes, with only a surplus of peaches for the non-storable fruit crops.

For storable vegetable crops, urban fringe agriculture in the GGH produces more than the annual requirements for the 9.1 million residents of the GGH for crops such as beets, carrots, and potatoes (see the last column in Table 12.1). In the case of storable fruits (apples), the region supplies 84% of annual requirements.

Fruit and vegetable production in this urban fringe does not meet all of the residents' requirements in a summer season, as is the case of strawberries with only 16% of the annual requirements supplied locally. At the same time, the region has production that is surplus to annual requirements, such as peaches, which implies the GGH specializes in peach production and ships fresh peaches to other parts of the province and into other provincial markets (such as Quebec). In the case of asparagus, the GH only supplies 6% of in-season requirements; however by expanding the supply area to the GGH the supply base provides for over 300% of summer time market requirements and 86% of annual consumption requirements for 9.1 million consumers.

Other examples of specialization in fruit and vegetable production and the GGH supplying other markets with fresh product includes beets, carrots, potatoes, and sweet corn (using the last column in Table 12.1 where the ratio is greater than 100%).

There are clear advantages to entrepreneurs that choose to engage in peri-urban farming and supply fresh produce. Such advantages include (1) proximity to the large consumer market in the metropolis, which is over seven million in the case of the GTHA, (2) low transportation costs from the urban fringe into the city center and the lower carbon footprint associated with the close proximity, (3) access to a nearby labor supply required for peri-urban farming that specializes in fruit and vegetable production, (4) ability to meets green city demands for fresh high quality foods, particularly fresh fruits and vegetables, and (5) the higher returns available to farming operations from supplying these green city requirements compared to more traditional farming operations.

Peri-Urban Farming Specializes to Focus on Green City Needs

Peri-urban farming in many parts of North America is in a transition stage, where peri-urban farming is focused on supplying the fresh foods (i.e., mostly fruits and vegetables) demand of the metropolis that is replacing traditional farming, which focused on either livestock production or grains and oilseed production.

Most grain and oilseed crops require significant processing prior to being presented to consumers as typical consumer ready food items purchased at retail, such as bakery products, cereals, etc., which are also shelf stable items. The economics of growing these crops require significant scale economies, which lend themselves to being produced in regions farther away from the urban fringe where successful farm operations can operate farms with a few thousand acres—this is difficult to achieve in the urban fringe. The size of equipment required for a successful grain and oilseed operation can conflict with traffic in the urban fringe. Ideally, the best and highest value use of farmland in the urban fringe is in directly supplying the needs of metropolis with, for example, fresh fruits and vegetables that require a much smaller land base than grain and oilseed operations (and fewer acres required to support a family operation).

Livestock production, when it is land-extensive is also best suited to areas outside of the urban fringe, when just considering the opportunity cost of land and using land in the urban fringe to its highest value and best use. At the same time, farms with livestock can be part of the agri-tourism supply base in the urban fringe. Large cattle operations, whether large dairy farms and large feedlots, exhibit better economics when located in agricultural areas beyond the urban fringe. Intensive livestock production in the urban fringe can present nuisance problems such as odors and noise pollution to nearby residents.

The highest and best use of agricultural land is in production of fresh fruit and vegetable production that meets the requirements of the metropolis, which is peri-urban agriculture. Peri-urban agriculture can include field production of fruits and vegetables, greenhouse production, and as well vertical farms. The latter can also be part of urban agriculture, and enclosed agriculture (e.g., greenhouses and vertical farms) can also occur when the soil and climate is less suited to field production.

Urban Fringe Agriculture Will be in Transition for a Period of Time

At the same time, traditional livestock operations and grain and oilseed operations continue to exist in the urban fringe. Transition from these more traditional operations to peri-urban farming that are focused on the needs of the green city can take time.

In the transition phase, which can take decades, traditional farm operations can remain in the urban fringe simply due to some family farm operations not wanting to switch from a cash crop or livestock business to supplying fruits and vegetables into the metropolis. These farms will only become part of peri-urban farming when the next generation of the family farm decides to switch its focus, or when the family farm is sold to an entrepreneur wanting to participate in peri-urban farming. Some farms remain in families and are rented out when there is no family succession—this can create an opportunity for entrepreneurs to enter into longer term lease arrangements that are conducive to investing in a fruit or vegetable farm operation.

An on-going challenge for peri-urban agriculture is that when farmland is listed for sale, in many cases, this land can also be purchased and used for pleasure such as hobby farms, horse farms, or just large country homesteads. Without zoning or land use restrictions, farmland in the urban fringe can quickly become part of urban sprawl or part of an industrial/commercial complex. Even when land is zoned as agricultural, which limits its use for commercial and industrial purposes in the urban fringe, non-farmer buyers can often place a higher value on farmland in the urban fringe than can farmers wanting to engage in peri-urban farming. This reality can limit the potential of peri-urban agriculture.

Land Use Policies Support the Potential of Peri-Urban Agriculture

There are significant demands for the farmland in the urban fringe where retiring farm operations will, in most cases, want to sell their lands to the highest bidder, which typically includes the best use of the land given the local zoning regulations and land use restrictions. If there are few zoning restrictions, the highest and best use in such an environment can be based on a developer's use of the land for a housing subdivision or for other nonfarm commercial enterprises. This works against development of peri-urban agriculture in the urban fringe.

Greenbelts around a metropolis, which restrict certain types of land use and/or restrict changes in zoning from agriculture to other uses, are most likely necessary to enable a significant amount of peri-urban farming. In the absence of such greenbelt policies, the highest and best use of land in the urban fringe is typically more urban development, which is at the expense of peri-urban agriculture. In Ontario, a significant portion of the GH includes the Greenbelt, which is the green shared area in Fig. 12.2 above. The Greenbelt surrounds the western shores of Lake Ontario, from Niagara-on-the-Lake, through the GTHA, to parts of Northumberland County. The purpose of the Greenbelt Plan is to permanently protect farmland and support the agricultural system, as well as the natural system and features within it.

The vision for the Greenbelt as noted in the Greenbelt Plan (2017)[3] indicates that:

The Greenbelt is a broad band of permanently protected land which:

- *Protects against the loss and fragmentation of the agricultural land base and supports agriculture as the predominant land use;*
- *Gives permanent protection to the natural heritage and water resource systems that sustain ecological and human health and that form the environmental framework around which major urbanization in south-central Ontario will be organized;*
- *Provides for a diverse range of economic and social activities associated with rural communities, agriculture, tourism, recreation and resource uses; and.*
- *Builds resilience to and mitigates clime change.*

The Greenbelt Plan also articulates goals, including those for *agricultural viability and protection*, which are:

- *"Protection of the specialty crop area land base while allowing supportive infrastructure and value added uses necessary for sustainable agricultural uses and activities;*
- *Support for the unique nature of specialty crop areas as our vital fruit and vegetable growing areas which include:*

[3] Greenbelt Plan (2017), which was approved under the *Greenbelt Act*, 2005.

- *The Niagara Peninsula specialty crop area, a destination for and center of agriculture focused on the agri-food sector and agri-tourism related to grape and tender fruit production; and*
- *The Holland Marsh specialty crop area, a centre of agriculture focused on the agri-food sector related to vegetable production;*

- *Protection of prime agricultural areas by preventing further fragmentation and loss of the agricultural land base caused by lot creation and the re-designation of prime agricultural areas;*
- *Provision of the appropriate flexibility to allow for agriculture, agriculture-related and on-farm diversified use, normal farm practices and an evolving agricultural and rural economy;*
- *Increasing certainty for the agricultural sector to foster long-term investment in the agri-food network and improvement to and management of the agricultural land base; and.*
- *Enhancing strengths of the Agricultural System, including through considerations for the impacts of development on agriculture and planning for local food and near-urban agriculture.*"

The Greenbelt has approximately 300,000 ha of farmland, along with the natural advantage of being located in the urban fringe of the GTHA. There also some natural advantages in certain areas such as former marshland where a number of vegetable crops are grown and the Niagara region, which produces most of Ontario's tender fruit crops (along with the Norfolk region). There is significant opportunity for more peri-urban agriculture in this Greenbelt area, as well as beyond its boundary, to provide fresh high-quality food to the metropolis of at least 7 million residents within the GTHA.

Chapter 13
Global Impacts of the Ukraine Invasion: A Resource Perspective

Leo Wambersie, David Lin, and Mathis Wackernagel

Abstract Ukraine contains 2.7 global hectares of bio-capacity per person, of which, nearly two (1.9 global hectares per person) are cropland. This is about four times the world average. Ukraine, however, only uses one-third of its crops produced domestically, with cropland dominating Ukraine's exports. When measured in Ecological Footprints, Europe is the largest consumer of goods from Ukraine, followed by the Middle East. For crops exported from Ukraine, the Middle East is a larger consumer than Europe. Here, we discuss potential implications the conflict in Ukraine may have on resource security around the world.

Keywords Global food security · Ecological footprint · Resource security · Ukraine

The Ukraine invasion comes with heart-breaking human suffering. It will also impact people far away from the war as global supply chains for food, metals, and energy are crippled. The price shocks they cause may well lead to deadly consequences, particularly in lowest-income countries, with the potential to amplify conflict and migration pressure (Robinson et al., 2022). This dynamic is one of the latest examples of the challenges highlighted in our research on "The importance of resource security for poverty eradication" (Wackernagel et al., 2019, 2021), which shows that 72% of the world population live in countries that ***both*** have a bio-capacity deficit ***and*** less than the world average per capita income. This means their demand exceeds what their own ecosystems can regenerate ***and*** they do not have sufficient income to acquire reliably needed resources from international markets.

L. Wambersie (✉) · D. Lin · M. Wackernagel
Global Footprint Network – California, Oakland, CA, USA
e-mail: leo.wambersie@footprintnetwork.org

A. A. Kubursi et al. (eds.), *Urban Food Security in a Crisis Prone World*,
https://doi.org/10.1007/978-3-031-89440-4_13

Given climate change challenges, biodiversity loss, freshwater shortage, deforestation, soil degradation, desertification, and food crises, it is becoming increasingly evident that the most limiting physical factor for humanity is not the nonrenewable stocks, but rather the renewable, in particular, biological resources that are massively overused. For example, fossil fuel use is more limited by the biosphere's ability to sequester carbon than by how much can be found underground. Similarly, access to minerals and ores is more constrained by the energy and biological resources needed to extract these resources rather than the total existing amount. Yet, when adding up all competing demands for the renewable resources of the biosphere, the results show that humanity's consumption demand now exceeds the Earth's regeneration by 75% (York University, 2022; Lin et al., 2018, https://data.footprintnetwork.org). Yet, to preserve a good portion of biodiversity, E.O. Wilson suggested for humanity to only use half the planet's capacity (2016). This indicates a factor of three is the difference between the current biological metabolism of humanity (1.75 Earths).

Before going any further, it is important to highlight the bio-capacity situation of food production in the Russian Federation and Ukraine, both of which are biocapacity-rich countries. All the data points presented here represent the year of 2018, since this is the latest complete set of UN data, which feed the National Footprint and Bio-capacity Accounts 2022 edition. Here, we present a few key results on potential implications the war may have on resource security around the world.

For context, the Ukraine contains 2.7 global hectares of bio-capacity per person, of which, nearly two (1.9 global hectares per person) are cropland. This is about four times the world average. The details of the Ukrainian Footprint and bio-capacity are shown in Fig. 13.1. In contrast, the Russian Federation contains 6.7 global hectares of bio-capacity per person, of which only 1 global hectare is cropland demanded by countries. The food portion of the footprint and the bio-capacity follows a similar logic as the entire bio-capacity and footprint. But here we map out food flows to indicate the implications the disruption of renewable resources from Ukraine and Russia could have on the countries that depend on these flows. Given the fossil fuel inputs for agriculture, particularly fertilizers, the biological resource perspective demonstrates the linkages between energy and food, and the implications travel with various ripple effects through economies in ways that are not anticipated at first sight.

The two bars on the left of Fig. 13.1 show the bio-capacity of Ukraine. The first column depicts the biologically productive area of Ukraine in unadjusted hectares: just over 60 million hectares. Since its land is nearly twice as productive as the world average, per hectare, Ukraine's bio-capacity in global hectares is nearly double its size in hectares. The four columns on the right in Fig. 13.1 show Ukraine's Footprints. Among them, the production Footprint is the largest component. It represents the resources required to run the economy. In it, crop production stands out and carbon emissions, i.e., the carbon footprint of production, are also a large portion. The trade components show that Ukraine's net exports, particularly of crop products, are high. Therefore, Ukraine's consumption footprint (the rightmost

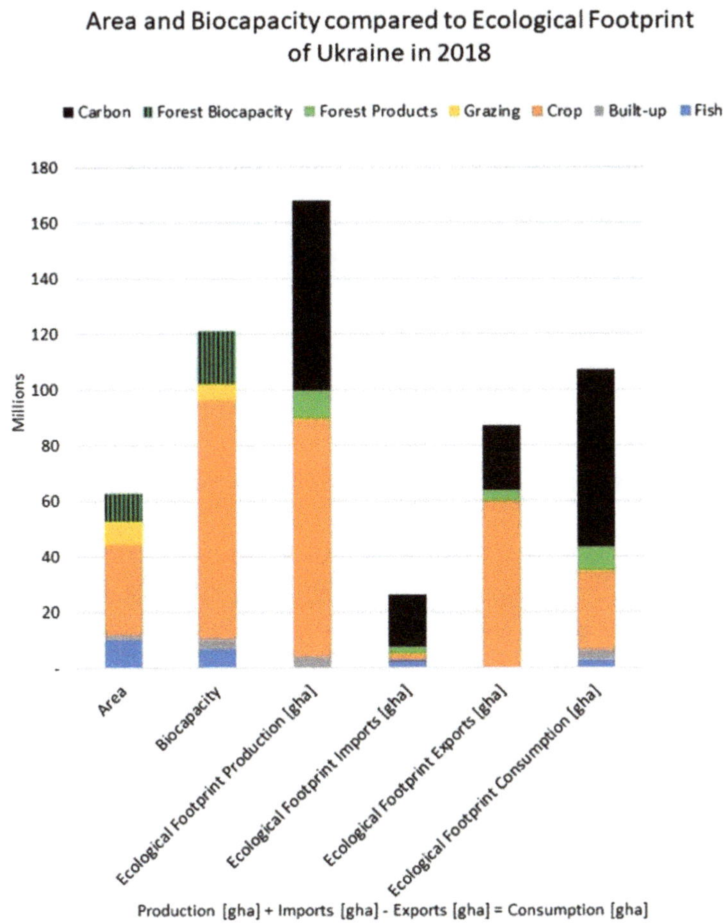

Fig. 13.1 Ukraine's ecological footprint and bio-capacity (2018)

column in Fig. 13.1) is much lower than its production footprint. For instance, its consumption crop footprint is merely one-third of its production crop Footprint, meaning that Ukraine only uses one-third of its crops produced domestically.

Figure 13.2 shows the contribution of both Ukraine and the Russian Federation in producing and exporting crops. Both countries are among the top ten producers and exporters of cropland bio-capacity.

Using environmentally extended Multiregional Input-Output assessments (ee-MRIO), it is possible to determine the final destination of exports, i.e., where the crop goods grown in Ukraine, as well as all other products produced in Ukraine, are finally consumed. Such an analysis is needed as general trade statistics only report "first port of entry" of exports, not their final destination. Given long supply chains, final products often travel through several countries. Figure 13.3a shows final destinations of crop products by country, Fig. 13.3b by region.

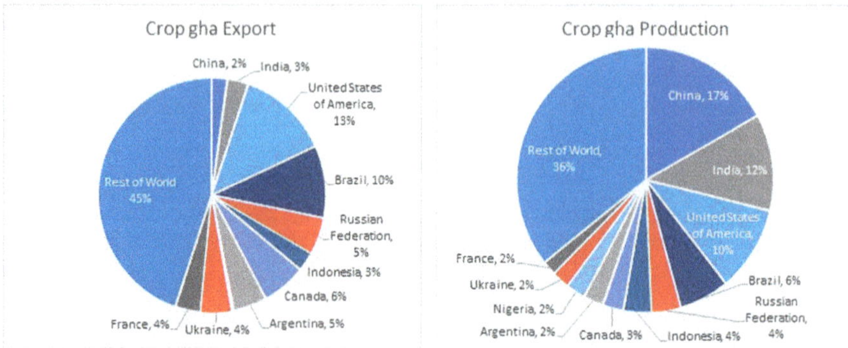

Fig. 13.2 Ukraine and the Russian Federation's position among top crop producers and exporters in the world (2018 data)

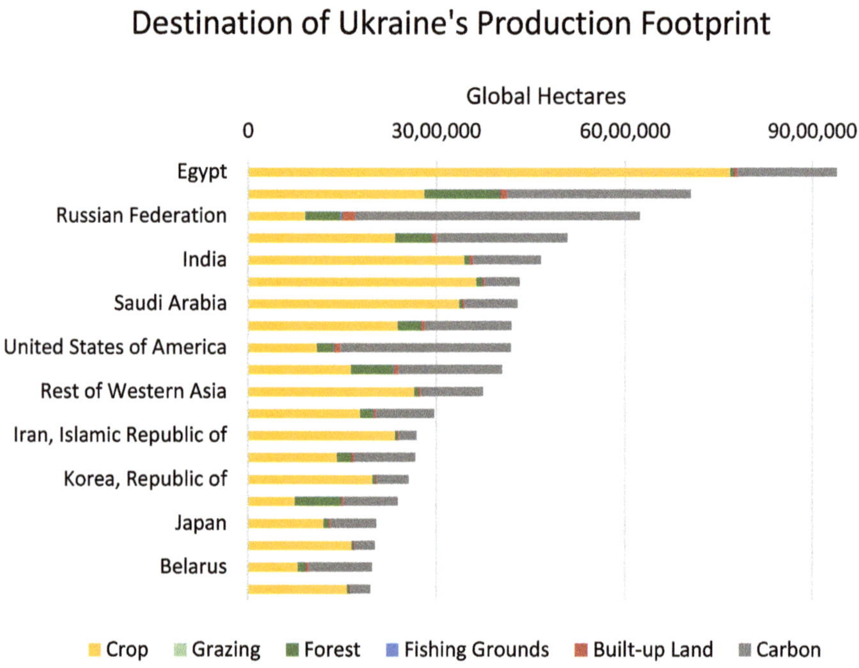

Fig. 13.3a Where production footprint of Ukraine is consumed by country in global hectares (top 20 in 2018)

Cropland dominates Ukraine's exports. For almost all the top consumers of Ukraine's exports, cropland is the largest imported footprint component from Ukraine. When measured in ecological footprints, Europe is the largest consumer of goods from Ukraine, followed by the Middle East. For crops exported from Ukraine, the Middle East is a larger consumer than Europe.

The footprint originating in the Russian Federation is much more varied as shown in Figs. 13.4a and 13.4b; including major forest products consumed in China,

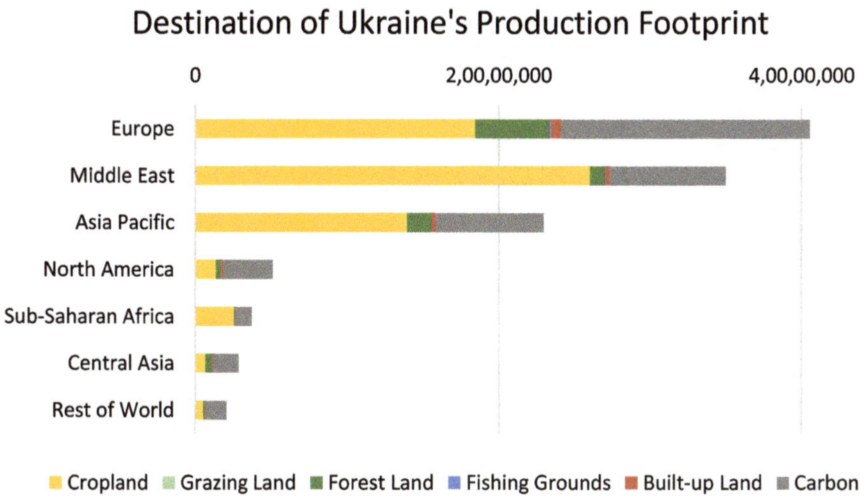

Fig. 13.3b Where production footprint of Ukraine is consumed by region in global hectares (in 2018)

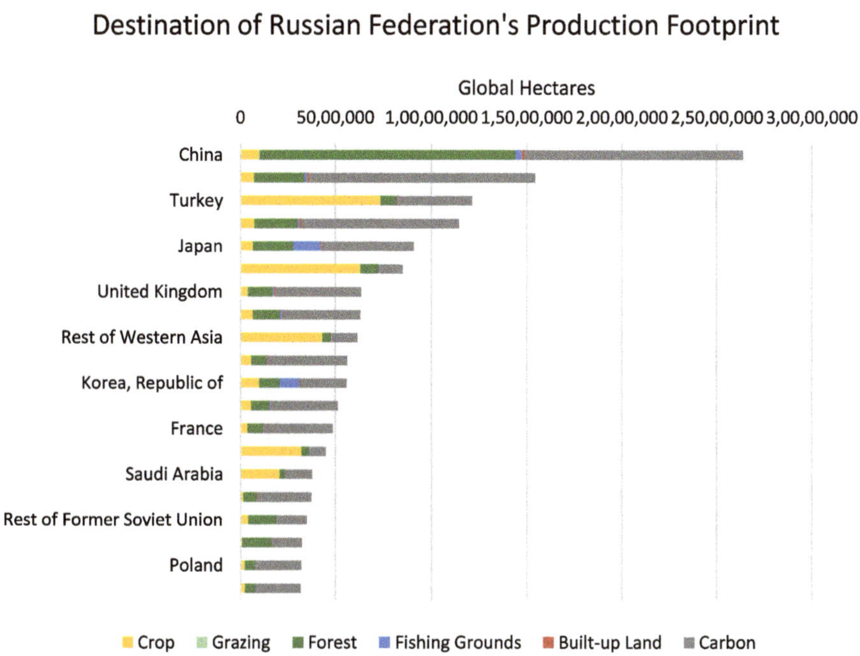

Fig. 13.4a Where the production footprint of the Russian Federation is consumed by country in global hectares (top 20 in 2018)

Fig. 13.4b Where production footprint of Russian Federation is consumed by region in global hectares (2018)

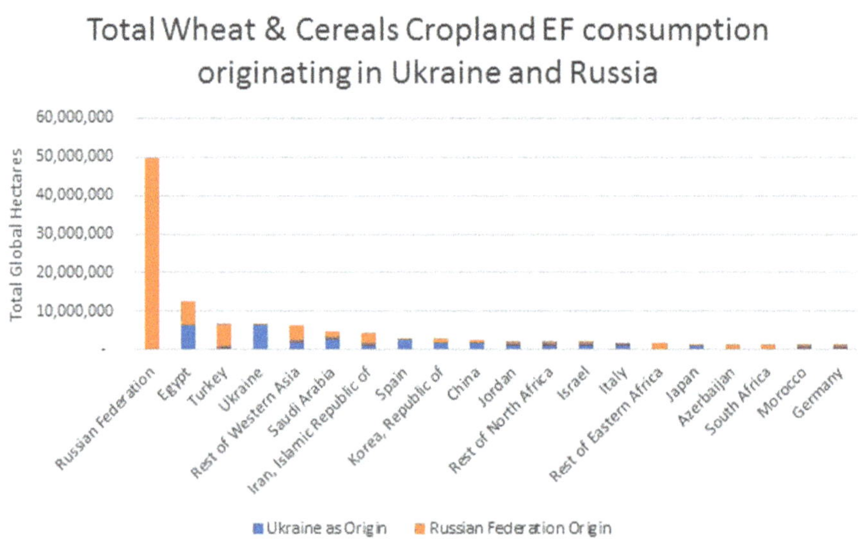

Fig. 13.5 Where Ukrainian and Russian wheat and cereals are finally consumed (in global hectares, 2018)

and embedded carbon footprint consumed in China and the USA. Interestingly, the Middle East is a major destination of cropland from the Russian Federation as well. Using the MRIO results, we can focus on the "Wheat and Cereals" sector of Ukraine and the Russian Federation (Fig. 13.5).

It is no surprise that most of the "wheat and cereals" production from the Russian Federation is consumed in the Russian Federation—and the same holds true for

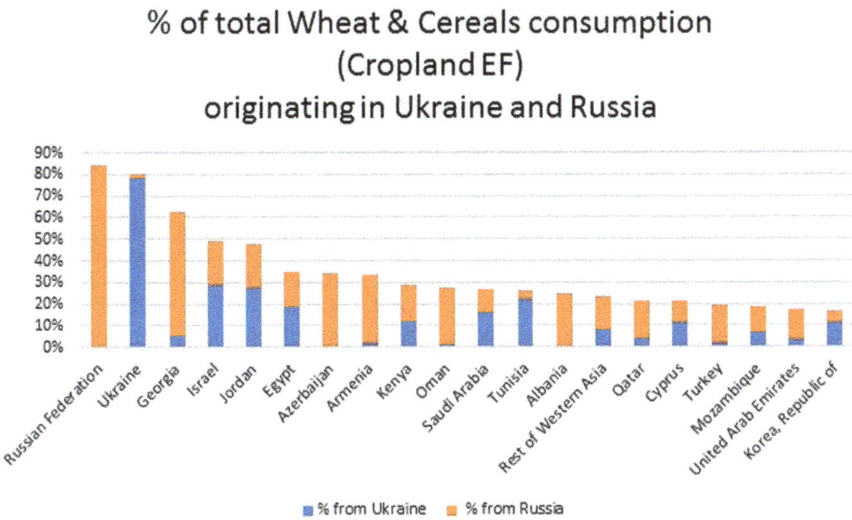

Fig. 13.6 Proportion of Ukrainian and Russian wheat and cereals in overall demand for wheat and cereals (in percent, 2018)

Ukraine. Egypt and Turkey, however, are large destinations of "wheat and cereal" that originated from both Ukraine and the Russian Federation. Even more revealing is the proportion of wheat and cereal consumed that originates from the Russian Federation and Ukraine (Fig. 13.6).

Aside from the source countries themselves, which we would expect to be the top consumers as well, the other top countries on this list are nearby neighboring Georgia and several MENA countries. In Tunisia, the country where the Arab Spring started in 2011, nearly one quarter of its consumed cereals and wheat originated in Ukraine.

Let us explain how one can interpret statistics like "over 60% of 'wheat and cereals consumed in Georgia originates from Ukraine and the Russian Federation" Does this mean that Georgia is irrevocably dependent on Ukraine and the Russian Federation? Looking at resource accounting data, as provided here, offers some insights. But it also brings up more relevant questions to consider when analyzing a country's resource dependency and determining associated risks. For instance, do these countries have price-competitive alternatives to Ukrainian crops? Given a lower level of price elasticity of basic food, could this significantly affect market prices, or even outprice large number of people from accessing basic calories? How will global markets and production systems all over the world adjust in the short and long term? Once we recognize how significant resource security is to provide for robust prosperity, these numbers help us answer some basic questions and formulate even more relevant ones.

References

Lin, D., Hanscom, L., Murthy, A., Galli, A., Evans, M., Neill, E., Mancini, M. S., Martindill, J., Medouar, F., Huang, S., & Wackernagel, M. (2018). Ecological footprint accounting for countries: Updates and results of the National Footprint Accounts, 2012–2018. *Resources, 7*(3), 58. https://doi.org/10.3390/resources7030058

Robinson, A., Csordas, V., & Wackernagel, M. (2022). Defining limits: Ecological overshoot as a driver of conflict. In O. Brown & G. Nicolucci-Altman (Eds.), *A white paper on the future of environmental peacebuilding, Geneva Peacebuilding Platform*. International Union for Conservation of Nature, PeaceNexus Foundation, Environmental Law Institute, Environmental Peacebuilding Association. www.ecosystemforpeace.org

Wackernagel, M., Hanscom, L., Jayasinghe, P., Lin, D., Murthy, A., Neill, E., & Raven, P. (2021). The importance of resource security for poverty eradication. *Nature Sustainability., 4*, 731–738. https://doi.org/10.1038/s41893-021-00708-4

Wackernagel, M., Lin, D., Evans, M., Hanscom, L., & Raven, P. (2019). Defying the footprint Oracle: Implications of country resource trends. *Sustainability, 11*(7), 2164. https://doi.org/10.3390/su11072164

York University Ecological Footprint Initiative & Global Footprint Network. National Footprint and Biocapacity Accounts, 2022 edition. Produced for the Footprint Data Foundation and distributed by Global Footprint Network. Available online at: https://data.footprintnetwork.org

Chapter 14
Urban Food Sustainability and Multilateral Environmental Governance Frameworks: Directional and Operational Complexities and Opportunities

Nidhi Nagabhatla, Gaurav Thapak, and Sisir Bhandari

Abstract The United Nations Economic Commission for Europe (UNECE) report, titled "People-Smart Sustainable Cities," highlights the intricate connections between urban development, the 2030 Transformation Agenda, and the Sustainable Development Goals (SDGs) and other Multilateral Environmental Governance Agendas (MEAs) like the Sendai Framework. The MEA outlines directional guidance, insights, and perspectives for program, policy, and regional/international cooperation that can be adopted or adapted toward sustainable urban development plans and policies. Progress has been made in implementing these global agendas, but significant gaps remain, particularly regarding food security for growing urban populations. In our synthesis, we tried deconstructing urban food sustainability, linking it to SDG 11, and advocating for resilient and sustainable cities.

This synthesis suggests that the urban plans and policies identify food security objectives and create concrete results-oriented activities plugged into multi-sectoral structures and interventions in the infrastructure, municipal, and community settings. We also emphasize the need for urban plans and policies to incorporate food security objectives within multi-sectoral frameworks. By using examples from various economies that demonstrate a commitment to integrating urban food sustainability into their obligations under the SDGs and the New Urban Agenda, this

N. Nagabhatla (✉)
United Nations University Institute on Comparative Regional Integration Studies (UNU CRIS), Brugge, Belgium

Faculty of Science, McMaster University, Hamilton, ON, Canada
e-mail: nnagabhatla@cris.unu.edu

G. Thapak
National Institute of Urban Affairs (NIUA) India, India, New Delhi, Delhi

S. Bhandari
United Nations University Institute on Comparative Regional Integration Studies (UNU CRIS), Brugge, Belgium

A. A. Kubursi et al. (eds.), *Urban Food Security in a Crisis Prone World*,
https://doi.org/10.1007/978-3-031-89440-4_14

chapter stresses the importance of recognizing interlinkages within the water-food-energy nexus and adopting innovative, nature-based solutions such as community gardens and urban food spaces. Furthermore, by analyzing current challenges and opportunities, we argue for a comprehensive approach to urban and peri-urban sustainability solutions (UPS) involving citizens, governments, and stakeholders in codeveloping effective responses and argue that the UPS vision requires alignment with existing sustainability.

Keywords Multilateral Environmental Governance Agendas (MEA) · Peri-Urban Sustainability Solutions (UPS) · Urban planning · Water-food-energy nexus

Introduction

Several factors, including climate change impacts, exacerbate the food security challenges and states and communities need focused attention on participatory mechanisms and inclusive policy support systems, particularly for vulnerable populations. For example, the IPCC Special Report on the impacts of global warming of 1.5 °C, in its summary on five integrative reasons for concern (RFC), classified indigenous habitats as "*unique and threatened systems, noting the limited capacity of these populations to address their vulnerabilities and community resilience*" (IPCC, 2018). This point reflects the disproportionate impact of stress factors on different populations and communities. Urbanization seems to be an obvious contemporary global phenomenon, and increasingly urban landscapes are driving economic and social development dynamics. This trend is affecting and pressuring (with emissions and industrialization) natural resources as demand for food, water, and energy supply is increasing, and is impacted by the climate change dynamics, such as floods and other climate extremes (Tollin et al., 2017). Overall, the multitude of stresses on water and land has threatened the sustainability of food production systems.

Development reports reflect on urbanization with quantitative information, a recent one stating that 2% of the global population lived in urban areas in 1900, while 50% in 2017, and projected to increase to 67% in 2050 (UN, 2019). The UN-Habitat forecast the growth of cities from 512 in 2016 to 662 in 2030 (Tollin et al., 2017). In addition, the production and consumption patterns are fast-changing and, currently, urban regions account for 56% of the total global population, making it a core factor in shaping the food supply and food security agendas at national, regional, and global levels (Buchholz, 2020).

The COVID-19 pandemic posed many and multifaceted challenges to the global economy and these set of challenges have directly and indirectly influenced (negatively) the water-food-energy nexus, more so for fragile economies and vulnerable communities (Ajnum & Nagabhatla, 2021) and urban and peri-urban settings hosting millions of people living in formal and informal habitats. The intensity of the pandemic affected economies, communities, and individuals at various levels, varying from country to country, however people in the low-income strata in any country witnessed cascading impacts, i.e., poverty and inequalities, and even an increase in barriers such as financial and human resource commitments toward the achievement of the Sustainable Development Goals (SDGs). For example, under the Agenda outlined in SDG 2 on Zero Hunger, it is projected that fulfilling related targets such as 2.1 that aim to end hunger and ensure food access to all people, in particular, the poor and the people living (including children) in vulnerable settings, may fall short because of the spilling effects of the COVID-19 crisis on national economies dealing with reduced revenue collections, tax immunity for the business that suffered huge losses, and financial support provided for the low-income groups or/and people with income loss (due to increasing unemployment in some countries).

Considering the stress that nations continue to face in addressing the challenges related to COVID-19, exploring alternative options to improve the production and distribution system for food and consumable goods remains crucial to achieving food sustainability goals. Note, for instance, target 2.4 on safeguarding sustainable food production (see Fig. 14.1 and Annexure 14.1 for additional dimensions constituting "urban and peri-urban sustainability solutions," hereafter referred as UPS). International, regional, and national experts, researchers, agencies, and institutions may have to play a collective role if in the remaining tenure (till 2030) of the SDG, achieving the outlined sustainability agenda while "building back better" and "leaving no one behind," is to be realized. Strengthening the structural dimension of food availability, accessibility, and affordability in the urban and peri-urban regions of the world is a central challenge in food sustainability planning and policies. Why? On account of the fact that a large number of people/population, in each country

Fig. 14.1 Keywords of the urban and peri-urban sustainability solutions (UPS) approach

lives in these areas (Alabaster & Nagabhatla, 2020). In order to further deconstruct the UPS approach, we divided the chapter into three sections.

Review of Recent International Reports/Concepts Aligning to UPS

The UNECE report "People-Smart Sustainable Cities" advocates a "cities-based" approach that recognizes the increasing role of cities and urban living in achieving sustainable development. As the urban areas host over 75% of the population in Europe, 80% in North America, and close to 50% in Central Asia, the report suggests the localization (urban scale) of SDGs and employing a "nexus"approach that combine multiple SDGs (3, 6, 7, 8, 9, 11, 12, 13) toward building "sustainable and smart cities for all ages" (UNECE, 2020). In recent years, increasing acknowledgments by the MEAs of significant societal trends and challenges associated with globalization, urbanization, climatic, and overall environmental challenges reflect an encouraging trend, although specific aspects such as urban food security are yet to be fully integrated into these agendas. The COVID-19 pandemic has shown how basic housing and planned urban development can contribute to public health, water, energy, and food security and underscored that urbanism principles should also take note of bottom-up approaches and inclusive strategies (Afrin, et al., 2021; Mouratidis & Yiannakou, 2021).

The "nexus"approach, in this chapter, incorporates "energy-food-waste-land" as interconnected systems and suggests looking beyond the traditional urban development and management paradigms. Sustainable cities foster innovative solutions, such as adopting the equitable principles of the UNECE Protocol on water and health and adoption of the nexus insights. Gondhalekar and Erlbeck (2021) reflect on the "Urban Nexus Development Cycle" as part of the nexus thinking and an innovative element toward guiding urban practitioners and overcoming "silo thinking." In this context, digital technology, where automation and AI (artificial intelligence tools) can support information aggregation, analysis, and applications for designing sustainable urban management strategies, plays a critical role. With food waste management a major concern for urban management and often an under-reflected aspect in urban development design, its link with food security goals and sustainable consumption practices is inevitable. Cities multiply the food loss problem because of their size, the constant need for food availability, and low cost for food. It is vital to enhance the understanding of food production systems and assess gaps in the value chains such as the lack of refrigerated storage and the low capacity for food waste management strategies. The UNECE Report also highlights an UPS strategy from New York, focusing on urban food waste reduction as a good practice example. Initiatives such as "United for Smart Sustainable Cities (U4SSC)" developed a set of key performance indicators (KPIs) to achieve SDGs. Over 100 cities worldwide are already implementing these KPIs by using computer information and communication technologies (ICT). Adopting the nexus approach, Switzerland, Lithuania, Denmark, and France also demonstrate best management practices on food waste to renewable energy. Reckoning that a city can act as a distribution point from the farm to the public while ensuring quality food and resources, this study also suggests cross-border and sectoral cooperation, city-focused cooperation, subnational approach to sustainable development, and integrating the urban system perspective in national government and policies are the key paradigms to shaping UPS approaches.

FAO's most recent synthesis on "State of the World Land and Water Resources for Food and Agriculture Systems at Breaking Point" reemphasizes the global population projections where the world population will increase from 7.7 billion in 2019 to 9.7 billion by 2050 (26% increase). It also points out that the fastest growth will be in low-income regions, including sub-Saharan Africa, where the population is expected to double by 2050 (FAO, 2021b). Uncontrolled urbanization shifts the socioeconomic variables of land and water resources and thus impacts agricultural production and consumption demand. By 2050, cities are expected to have two of three people living in less-developed or underserved settings (peri-urban, informal habitats), particularly in the regions of Asia and Africa. Noting that urban dwellers consume 80 percent of all food produced, urban development trends will directly drive food and socioeconomic security settings and the current tendency of people to move to urban spaces; these numbers reflect the immense challenges that cities/urban managers may have to embrace and address. In situations where the COVID-19 crisis has caused pushback to implementing multiple SDGs, refocusing attention on these challenges remains crucial for managing urban sustainability.

Another key point from this global report includes the shifting focus of FAO's commitment post the UN Food Systems Summit in 2021 toward inclusive land, water soil management goals, and the SDGs. It also acknowledges the rising urbanization rates and that expanding urban populations are and will increasingly displace agricultural land use and stimulate land-use changes in rural landscapes. In some regions, the development of urban centers and industrial sites is at the cost of drained agricultural landscapes, which leads to a detrimental impact on environmental flows such as an increase in flood runoff that in turn results in compromising the food security targets. Other challenges such as lack of capacity to manage agricultural waste are linked to source water pollution. Combined, these set of issues have significantly threatened food safety and food security efforts. While this report has a limited focus on food systems and food security in the urban context and SDGs (does not reflect any aspects of SDG 11 or sustainable urban development), other MEAs agendas are not considered in an integrated manner or largely overlooked. However, it reckons that international conventions and high-level political commitments provide a strong mandate and support for multisector and integrated land and water governance and achieving the SDGs.

Overview of Synergistic and Relational Connections Between Urban Development and Five Multilateral Environmental Governance (MEA) Agendas with a Focus on Food Security

Different MEA agendas emphasize the need for inclusive urban development; we provide below some key points from selected MEAs.

The 2030 Transformation Agenda: The Sustainable Development Goals (SDGs)

Approximately 1.3 billion tons of food waste is generated every year (UNEP, 2021) accounting for USD 680 billion and USD 310 billion from food waste in industrialized countries and developing countries, respectively (UNEP, undated). SGDs show interlinkages with food systems as noted in Goal 2 (zero hunger), Goal 11 (sustainable lifestyles and sustainable cities/communities), and Goal 12 (responsible consumption and production), and their agendas combine agriculture, food security, urban development with sustainability pathways calling for a low environmental footprint (Hoover, 2017). From the context of the UPS approach, SDGs provide a people-planet-centered perspective where understanding and analyzing the potential impacts and synergies is suggested as crucial (IIASA, 2018). For instance, SDG 11 calls for making cities and human settlements inclusive, safe, resilient, and sustainable and to enhance inclusive and sustainable urbanization. And the food-oriented agenda outlined in SDG2 calls for an end to hunger and achieving food security and improved nutrition along with the promotion of sustainable agriculture, ensuring sufficient food, and maintaining sustainable production and functioning of food commodity markets. These challenges in achieving the SDG 2 targets are massive and the responsibility for doing so should be taken beyond the conventional agencies and actors (rural farmers and fishers) in the food production sector. The realm of this responsibility needs to diversify and involve nonconventional stakeholders, such as urban dwellers, and utilize available urban landscapes as food production zones. Note that SDG 12 recommends sustainable consumption and production patterns, this objective has a direct bearing on UPS solutions, for instance, solutions addressing food waste reduction targets (that reads halving the global food waste by half and reducing food losses during production and supply chains, ensuring food security). Cities are home to billions and drive global markets, prices, and food security, managing through various innovative strategies; however, achieving this agenda could foster a new valuation in urban food sustainability pathways (Bellina, 2016). Overall, the SDG vision tackles and outlines key priority areas in ensuring food security, either directly or embedded in various food-sector-oriented goals. Box 14.1 provides an additional narrative to support the above outlined discourse.

The Ramsar (Wetland) Convention Wetlands (natural or constructed) can contribute to the conservation and restoration of biodiversity, terrestrial and marine ecosystems, and help secure food security and elevate the quality of life for the urban population (UNEP, 2020). Supporting this statement, the intergovernmental treaty "wetland convention" has encouraged governments since 1975 with a framework for national action and international cooperation to conserve and wisely use wetlands and their resources (Ramsar, 2016). The overarching objectives of the convention also address the sustainable development agenda by suggesting the need to safeguard biodiversity and wetlands. Recognizing that rapid urbanization and climate change challenges ongoing efforts of states/nations to meet the food security targets, MEAs like the wetland convention have enhanced focus on urban landscapes/wetlands including support for interventions that enhance the management of water and air quality, disaster risk reduction, and climate change adaptation (Nagabhatla & Metcalfe, 2018). The Convention seeks to establish interagency support in cross-cutting issues while supporting the urban wetlands concept including a focus on green infrastructure, nature-based solutions, and related socioeconomic agendas such as improving the quality of life, human and ecosystems health and well-being. The 2013 report, by the convention stressed the connection of sustainable urban development with food security suggesting that the wetlands should be considered solution providers that mitigate risks from climate change and support food production to support the needs (including income generation through recreation and tourism activities) of the growing population. Over years, the Convention has moved beyond conveying the wise management of wetlands message to widen its scope and highlight the importance of urban wetlands and interlinkages between wetlands, freshwater systems, urban planning, and sustainable development (Finlayson et al., 2013). In 2018, the World Wetlands Day has themed "Wetlands for a Sustainable Urban Future."

Box 14.1 Shifting Paradigms (and Impact Domains) in the Food Security Agenda

India's Green Revolution of 1960s was a crucial step toward food self-sufficient. While the foundations relied heavily on mechanization of the agriculture sector, use of technology and genetics to optimize the yield with emphasis on rice and wheat, use of fertilizers, irrigation supplies etc., resulted in a threefold increase in production in four decades (John & Babu, 2021; Singh, 2000), the area under cultivation increased only slightly. Evaluation of this paradigm reflects differentiated impacts besides the fulfillment of food security goals such as loss of indigenous knowledge systems for food production and fast/vast changes in the nutrient composition of the soil (Eliazer Nelson, et al., 2019), soil degradation, erosion, and compaction and in depletion of groundwater table and quality. Note that the increasing demand for irrigation was largely catered by groundwater use, and only about a third was provided by surface water systems (John & Babu, 2021; Singh, 2000). In tandem, the country witnessed high population rise in the decades that followed and recorded multifold increase in water demand for the urban sector (Sadashivam & Tabassu, 2016; Singh, 2009). This conflict in water demand has long started to stress the water and land resources in the country, and climate change impacts exacerbates the cascading risks of water stress/crisis noted in different landscapes, including urban (Pingali, 2012). The "call of action" on integrated solutions has never been so urgent, and suggestions like UPS seem fitting.

The Paris Agreement (2015)

This MEA explicitly recognizes the "fundamental priority of safeguarding food security" and "sustainable lifestyles and sustainable patterns of consumption and production" where "parties take the lead" and "play an important role in addressing climate change" (UN, 2015; UNFCCC, 2016). It binds countries to have Nationally Determined Contributions (NDCs), strategies that reflect on national efforts toward sustainable development (at least or over six SDGs including 2, 9, 11 & 14), and climate change mitigation goals (Dzebo et al., 2019). For instance, in climate-smart agriculture planning, strategies with a similar configuration as UPS find a clear mention. It connected SDG 2 to the third-largest share (13% of the total) of NDC activities; however, there is no clear link between food security and urban development. Progressively, countries underline the NDC as a key policy document and some have prioritized mapping NDCs and SDGs overlaps. In most countries, NDCs attribute human activities primarily responsible for climate change; the adaptation priorities generically and specifically refer to the urban context including aspects of food security, land use management, sustainable urban development, health, and improved governance, addressing vulnerabilities, disaster risk reduction, biodiversity/ecosystem conservation, and water management. Analysis of global NDCs, particularly in Africa and Asia shows that food security is fairly represented in adaptation planning, for example, in Costa Rica, NDCs promote renewable energy and integral environment management through agroforestry systems, watershed management, and municipal urban planning (Tollin et al., 2017) along with enhanced support for land-use policies supporting food security and infrastructural resilience. While the Paris agreement suggests relevance to the urban context, systematic attempts to "integrate climate policy objectives into the urban dimension of cohesion policy" are still a work in progress (Nesbit, et al., 2017). Regional committees support the Paris Agreement toward promoting integrated sustainable urban development policies across all levels, however, food security with sustainable consumption and management perspective context is largely missing. We acknowledge that better integration of UPS in the NDCs and climate change adaptation/management strategies (RUAF, 2020) holds promise.

The New Urban Agenda

Global reports predicted the enlargement of cities and how the urban landscape will have to prepare to host additional 3 billion inhabitants by 2050 (Clos, 2015). The New Urban Agenda (NUA), unanimously adopted by states, outlines a vision for sustainable urban development aligning with the SDGs (Kalapurakal, 2018). Reckoning that urbanization is the most transformative trend of the twenty-first century, addressing food security for the cities, population, economy, and society is both challenging and a high priority task. Cities and human settlements are active dimensions of the social and ecological functioning of urban landscapes. NUA strategic documents mention the context to food nearly 26 times where the focus is rendered on well-connected and well-distributed networks of production and supply chains (UN, 2017). The agenda also casts a road map for boosting local provisioning of goods and essential services that can strengthen the sustainable management of resources, including improving food security. The agenda points to well-connected green and quality public spaces to improve urban sustainability, including but not limited to improved strategies for waste management, food waste and wastewater management. NUA holds the potential to support the UPS framework as urban institutions and territorial planning are progressing, and the role of subnational entities like cities and towns in MEAs is noted. UNDP provided full support to implement the New Urban Agenda with the official launch of its Sustainable Urbanization Strategy (Xu & Tuts, 2021), highlighting that UPS solutions also could enhance the role of cities in mitigating poverty and improving livelihoods by agreeing to the energy-food-water-waste-land nexus; however, the explicit mention of urban food security perspective is missing (UNECE, 2020).

The Sendai Framework

In 2015, the third UN Disaster Risk Reduction (DRR) Conference in Sendai, Japan, adopted a nonbinding global agreement to reduce disaster risks, the Sendai Framework for Disaster Risk Reduction, to promote a shift from "disaster response to disaster risk reduction and management" while developing infrastructures and human settlements immune to disasters systematically linking the framework with SDGs and other MEAs (IIED, 2016). Between 2008 and 2018, the impacts of disasters imposed a cost on the agricultural sectors of developing countries that exceeded USD 108 billion in damages or lost crop and livestock production (FAO, 2021c). Noting an average of USD 170 billion economic losses from disasters, the Sendai Framework is helping bridge policies' gaps to buffer disaster impacts in the agriculture and food sector (ibid), including seeking cross-border collaboration and coherent implementation of disaster management tools and plans. This Framework prioritizes action concerning food/food security and UPS are noted multiple times, and concerns around consequences of unplanned and rapid urbanization are well-reflected in the action plans. Calling for the establishment of scientific mechanisms to address land use and urban planning and designing land-use policies in urban settings, the framework stresses that the role of cities in terms of local government roles and responsibilities, particularly for DRR, should be seriously viewed (IIED, 2016). As countries are implementing the Sendai framework, cities and urban governance could integrate the food security agenda in these plans and interventions to achieve SDG 11. An example, "Making Cities Resilience" is a global campaign to increase understanding and encourage the commitment of DRR by local and national governments (Eltinay & Harvey, 2019). It acknowledges the impact of medium and large-scale disasters, including in the food value chain, with negative consequences on trade flows of agricultural commodities. FAO identified the interconnection between SDGs, Sendai Framework, and Universal Climate Change Agreement, such as climate change to address food security and sustainable urban development in many of its synthesis reports (FAO, 2015). The Sendai Framework has renewed international commitments and reflects an enhanced framework that galvanizes and reinforces efforts to mainstream risk reduction across the agriculture sector, predominantly given its core outcome: "the substantial reduction of disaster risk and losses in livelihoods" and also in "businesses, communities and countries" (Eltinay & Harvey, 2019).

Case Study Approach to Illustrate Best Management Practices

This section showcases selected case studies to expand on the descriptions from Sections "Review of recent international reports/concepts aligning to UPS" and "Overview of synergistic and relational connections between urban development and five Multilateral Environmental Governance (MEA) agendas with a focus on food security". As seen in Table 14.1, we select five examples where UPS framework objectives converge with the smart cities' vision.

The discussion below is about an integrated perspective toward food-secure futures for urban communities, referring to the framings and guidelines of the MEAs and points to the UPS framework as a constructive approach that bears the capacity to address multifaceted environmental and climate change impacts. We also emphasize that such approaches should appropriately connect with local and traditional knowledge systems, address context-based norms, and integrate aspects of social justice, water, and food security targets specific to that setting. The ten key elements of UPS (Annexure 1) offer a checklist to create an integrated vision that can boost the integration of the MEA agendas in urban planning and programs. Sauvé et al., (2021), argue that an intra-sectoral and cross-sectoral application of technology and innovation can address the challenges in resource management, pointing to the emerging concepts of circularity in the water domain. Further, Avellán et al., (2022), underline that challenges related to climate and water crisis, vulnerability, and resilience gaps are many, therefore, solutions that are designed with innovative mechanisms/approaches like a circular economy can also offer

Table 14.1 Selected case studies to expand on the emerging paradigms of UPS

Case study 1: India: Integrated production systems in urban wetlands of eastern Kolkata Wetlands of eastern Kolkata spread nearly 12,500 ha, of which 32% is used (mainly ponds) for fish farming and 39% for agriculture. The wetland receives water along with sewage from the city, mainly from two canals and treats about 600 million liters of sewage per day. The sewage is received in 286 ponds that are used for commercial fish production. The depth of the ponds varies generally between 50 cm and 150 cm. The shallow ponds are efficient in the primary treatment of sewage that eventually feeds the wetland. The sewage provides essential nutrients for the fish production, and the farmers have an established system of treating the ponds and wetlands to maintain water quality, soil fertility, and designing fish and vegetables production systems with fish species from both indigenous and exotic varieties ((Bhattacharyya, 2021; Vicziany, et al., 2017). While this design fits with the major structural guidelines of UPS, in scaling such water and soil treatment systems, integration of local and traditional knowledge from fishers and farmers is a key aspect. Various researchers have documented the benefits and reduction of pathogens through the processes deployed by the farmers in the wetlands (Bhattacharyya, 2021; Vicziany, et al., 2017). The farmers also harvest vegetables on the ponds after emptying them for drying periodically. The wetland ecosystems provide not only ecological benefits of treating wastewater from the urban landscape, but this solution also allows for protecting local and native biodiversity while adding economic co-benefits of providing livelihood to farmers and contributing a good share of the urban food demand as noted, i.e., nearly 150 tons of vegetables and about 35 tons of fish every day. While the city and the farmers rely on the wetland for the produce, overall, the ecosystem services from the wetland and the urban landscapes benefits from such sustainability practices adopted by the local communities and citizens (Rekow, 2020; Vicziany, et al., 2017).
Case study 2: Thailand: On rehabilitating dumpsite as a farmland The conversion of landfills to an urban farm in Chiang Mai, a sprawling urban center in Thailand, reflects an innovative UPS strategy. This city in northern Thailand is largely dependent on tourism; however, for the past almost 2 years, the COVID restrictions (lockdowns and other measures) has caused severe economic loss for the residents and the local economy. Loss of livelihood opportunities due to the pandemic exposed the vulnerability of citizens and other residents purchasing capacity for daily provisions. In response, families of low-income groups in the city designed a community architectural intervention, i.e., transforming a landfill in the city into an urban farm, the city had marked the landfill as "un-used" with the plan to utilize it for tree plantation and alternative housing projects. However, the COVID crisis triggered a response from the community to create an urban farm, working together with the community, the city administration cleared about 5700 tons of waste from the landfill, spreading over half a hectare. Post the leveling procedure, the topsoil was replenished to prepare for the intervention. The funds for the project were crowdsourced from donations, and the human resource was drawn from the community and the municipality. Currently the community grows a variety of grains and vegetables. Along with the farm, a small bamboo structure is developed that is used for meetings, training, and as farmers' market. Chiang Mai has presented the best example of the success of decentralization and a self-reliant approach to urban food demand management (Boonmahathanakorn, 2020). During the pandemic, the city also noted the emerging consciousness in people to utilize their backyards and other open spaces to produce vegetables and fulfill their food needs; however, in few cases public space was designated for farming and agriculture with support of the local authorities (Boonmahathanakorn, 2020; Chandran, 2021).

(continued)

Table 14.1 (continued)

Case study 3: Serbia: *Belgrade exploring options for multiple use and benefits of organized urban waste management (focus on landfills)* The intervention for remediation of air and water contamination in Vinča landfill in Belgrade, Serbia, is another example of the state's machinery to embrace sustainable development guiding principles in the urban planning context. This landfill in Belgrade, located about 12 km away from the city center on the banks of the river Danube, is the largest unmanaged open dumpsite in Europe. The site has received about 3000 tonnes of construction waste and about half that quantity of domestic waste per day for more than four decades. The leachate from the waste has been contaminating the surface water of Danube and groundwater in the region as well as the toxic emissions from the unmanaged waste pollutes the air. In 2019, a project to transform the dumpsite into a sanitary landfill was initiated to create safe management of solid waste from the city and treat the existing waste of the dumpsite for fuel and energy production. The World Bank supported intervention led to the creation of a new facility for the conversion of waste to energy, the facility is still under construction, along with a new space for a sanitary landfill, and an incineration plant (Mayhew, 2020). It is anticipated that landfill remediation will prevent many potential future hazards while creating a space for safe and sustainable urban waste management. The UPS design takes note to address the harmful methane emissions, thereby reducing exposure to multiple risks, including fire hazards, a common occurrence for the site. The proposal also involves plans to collect rainwater, industrial effluents, and leachates and appropriately treat the wastewater to maintain the soil and subsurface water quality in the region. In this case, the benefit of water quality management is directly applied to the farmers in the Danube basin that depend on the river for irrigation (European Bank, 2018; Mayhew, 2020; Vasovic, 2021).
Case study 4: Mexico: Employing energy security to channel water and food security obligations Mexico plans to achieve sustainability goals through engagement in the energy sector, the intention is to reduce emissions from the agriculture sector and achieve multiple SDGs. To this vision, the Government of Mexico, the World Bank, and the Global Environment Facility have collaborated to promote environmentally sustainable technologies while identifying the need that energy is a significant requirement in the small and medium-sized businesses in the agriculture sector. Agriculture itself was noted to contribute about 7% in 2007 (INE, 2007) and 12% in 2015, of the national greenhouse gas emissions (INECC & SEMARNAT, 2015). Through Fideicomiso de Riesgo Compartido (FIRCO), the Government of Mexico invited proposals from small and medium-sized farming and agribusinesses to develop business models and integrate environment-friendly technologies in the energy sector. Through the support in legislation, the businesses also moved on to sell energy to the energy grid of the country. The FIRCO promoted efficient energy systems and technologies for renewable energy. With the growing impact, scaling these systems from the provision of renewable energy technologies to promoting technologies for reducing the use of synthetic fertilizers, restoring soil fertility, and reducing the emissions emanating from the mishandling and decomposition of organic matter is the transformation that fits with an integration agenda perspectives/framework. The initiative has simulated 2286 renewable and energy-efficient technologies adopted by 1842 small and mid-sized agribusinesses and contributed to the reduction of >6 million tonnes of CO_2 emissions and 382 million kWh of energy saved. Additional benefits include earnings for farmers from the sale of energy, increased productivity of agriculture, and reduced carbon footprint. The case of Mexico also demonstrates the integrated thinking spearheaded by MEAs and the agenda obligations that the states have endorsed wherein the national governments and the citizens are engaging to fulfill/implement sustainability goals and targets in multiple sectors, including but not limited to water, food, and energy security "for all" (FAO, 2021a; World Bank, 2019).

Table 14.1 (continued)

Case study 5: South Africa: Support for urban Agriculture on the policy radar In the Gauteng province of South Africa, where the megacity of Johannesburg is located, urban agriculture is supported through policy for a resilient and equitable urban development. The province is developing an urban farming policy to support and provide a framework for the city farmers and residents keen to cultivate food systems (Cilliers et al., 2020). The objective of this policy is to provide support and help develop enabling environments for farming and agriculture at all levels of urban life: home, community, and larger commercial farming. Researchers have acknowledged that the success of any UPS strategy in the city is a result of the proactive policy measures and innovative mechanisms such as "extended social packages" and "food empowerment zones." The citizens actively promote urban farming practices in the city through rooftop hydroponic farming in the inner city (Cilliers et al., 2020; Steenkamp et al., 2021). The Urban Agriculture Initiative (UAI) supports multiple vulnerable and young citizens in their entrepreneurial endeavors toward urban farming through finances, capacity building, and training. Alongside, the Department of Health and Social Development established a pisciculture facility in the city using a production model that utilizes less water and non-pesticide-based farming techniques and produces about 2 tons of vegetables per year catering to the food demand of the neighborhood and the surrounding areas with a high level of poverty and unemployment (Cilliers et al., 2020). The city of Johannesburg is a good example to showcase the UPS model for urban agriculture to realize food security and community resilience. However, the success of the model is largely in its participatory approach that proactively acknowledges multiple issues and supports a multipronged solution.

alternative viewpoints and proposes that mainstreaming the agenda of "zero waste" and "smart management" and creation of decision support systems at the local and national levels for the communities and populations has scope to provide incentives for involved stakeholders. Also examined are concepts like "smart cities" (Section "Case study approach to illustrate best management practices") to reflect on the practices that are applied at the local scale and how these examples can serve as a best practice reference. Overall, the synthesis aims to bridge the concepts of food security, water security, and climate security future by proposing an enhanced focus on "integration," "collaboration," and "collective action" toward designing urban resilience, i.e., UPS strategies. In general, the urban sustainability plans created for a specific case (city or urban setting) can fittingly account for the theories, concepts, best management practices, and empirical understanding underlined in MEAs and global/region knowledge outcomes/reports.

Importantly, innovations in framings and application of the frameworks can support food security interventions in the urban settings by increasing awareness on food waste, separating food waste collections, introducing legislation controlling food waste, and capturing technologies that can generate energy and environmental advantages from the acquired waste (Jain et al., 2018). For example, the reduction and treatment of urban food waste is an effective way to reduce the carbon footprint of the city. In recent years, the growth of food delivery chains and ready-to-eat food in urban and peri-urban settings has changed the ways people produce waste. Jain et al. (2018), reflect how cities can ensure food security and sustainable development, referring to Hartberg, Austria, a town where food waste is collected and delivered to a biogas plant. The city provides regulations against food waste and segregation of waste aligning with SDGs. The food waste is taken to the biogas plant for addressing energy (heat and electricity generation needs), and farmers use the digestive for soil enrichment and increasing production. Similar approaches are reported from Columbia, Copenhagen-Denmark, Auckland-New Zealand, Milan, Italy, Minneapolis-USA, Oslo-Norway, and Seoul-Korea, and multilateral development in many other regions in the global north and south.

It is commonly acknowledged that SDG visioning has steered reforms in urban development strategies. Many examples are available to show that cities can engage in multipronged strategies and structural incentives to reduce waste and redistribute surplus foods. The Milan Urban Food Policy Pact is an example of how cities can develop sustainable food systems (MUFPP, 2020). This pact signed by 120 cities worldwide aims to develop sustainable food systems, integrate urban food policy considerations, and review and amend urban policies, plans, and regulations to establish equitable, resilient, and sustainable food systems globally. Increasingly, the MEAs have incorporated specific SDGs in their mandates, thereby progressively steering the food security and evidence-based management and policy support, including in and for urban and peri-urban settings.

Concluding Remarks

Considering the argument presented in the above sections, the complexities of applying UPS are diverse and multifaceted. The reference to the MEAs is a key point in designing UPS frameworks aligning with the agendas of sustainable urban development and food security "for all." In that context, MEAs focus on urban systems provides directional and operational guidelines and a proposal to cocreate urban sustainability-focused interventions that are inclusive and support the vision of "leaving no one behind." The discourses such as UPS are not only pertinent but necessary to offer knowledge for policy reforms, to collect and integrate multiple insights and create an integrated urban policy. In addition, outlining synergies and positioning multiple objectives to boost regional/international cooperation for sustainable urban landscapes is equally important, as good practices can be referred to and experiences shared. These new paradigms in shaping urban sustainability and food security are both important and timely. Moreover, the selected case studies address systemic inequalities embedded in resource allocations and policy discourse in the urban management landscape, as well as promise UPS interventions. We recommend closing the gaps and attending to the needs in practice/policy of urban management, including participatory implementation of food goals underlined in international agreements, MEAs, including the SDGs, and mainly SDG 11: Sustainable Cities and Communities.

As per conventional thinking, food production has been associated with rural landscapes and labeled as the responsibility of farmers and fishermen to ensure that all food demands are met. However, as the implementation of these MEAs is in progress, it points to the fact that food production-consumption dynamics for the burgeoning population is far below the scale needed, and the food security agenda should be diversified. Intentionally, the reference to National Sustainable Development Strategies (NSDS) proposed in 1992 in Agenda 21 is drawn in order to support the narrative. This is because nearly 30 years ago, this agenda had outlined the need of reconnecting with inclusivity-oriented guiding approaches wherein the states acknowledged the integration of economic, social, and environmental objectives into one strategically focused blueprint for action. NSDS reckoned that integration through the widest possible participation is based on a thorough assessment of the current situation and initiatives, and by the next decade (2002), only about 85 countries have organized national strategies toward this vision in some form. Subsequently, the Johannesburg Plan of Implementation (paragraph 162 b) adopted at the 2002 World Summit on Sustainable Development (WSSD) steered nations to prioritize progress in NSDS. However, even after another two decades (2021), ownership and strong political commitment for MEAs are progressing at a slow pace. MEAs have long propagated integration across sectors, territories, and generations and the broad participation and effective partnerships, development of capacity and enabling environment, and mobilization of means of implementations.

While it is promising to see some cities/urban landscapes are prioritizing the sustainability agenda and designing local/context-specific UPS (see Section "Case study approach to illustrate best management practices" case studies), serious gaps and barriers are also reported. Future interventions must consider key gaps and needs toward integration of food security objectives in urban plans/policies and take note of best practices, innovations, projects, and programs in developed, developing, and emerging economies that have shown commitment to addressing urban food sustainability through technical, financial, and policy support systems. As countries and communities are targeting post-COVID recovery vision "building-back-better," a comprehensive overview of current and future challenges, as well as needs, can provide opportunities to plug the food security objectives in designing and implementing integrated urban policies. Besides, we suggest including a common set of indicators to assess the urban food dynamics while measuring and monitoring the progress of the MEAs in the urban settings. In conclusion, the synthesis by Ilieva (2016) points to "seeds of transition" in the context of shared urban food planning agenda as promising options in the direction of sustainable urban future pathways. And to close, to boost the operational efficiency of the proposed framework (UPS), communities and planners and all stakeholders should be actively engaged in codeveloping organized solutions to the multidimensional challenges in the urban sector.

Annex 14.1: Ten key Considerations for Designing a Context-Oriented UPS Framework

1. Creation of inclusive land and water management policies in urban planning, looking beyond conventional demand-supply dynamics.
2. Organize a coalition of informed and skilled urban experts, institutions, and organizations and adopt a multilevel governance system for effective and efficient options and operations.
3. Implementing food security-oriented UPS solutions at scale—neighborhood community, municipality.
4. Boost capacity of people and communities on technical skills to sustain the interventions and provide financial support systems and incentives for innovative technologies for contribution to urban food sustainability solutions.
5. Adopt best practices in the global food production knowledge base like the fertilizer code that guides against misuse, underuse, or overuse of fertilizers to create healthy and sustainable urban food production value chains.
6. Promote value-investing in UPS from the corporate sector/business community.
7. Empowers urban people/communities to be part of food security solutions (local/city scale).
8. Interoperable and scalable design, structures, and other dimensions.
9. Map the gaps and assess the need for designing integrated UPS systems. For instance, relatively underreported food waste management connection with food security.
10. Clearly defined incentive structures for better "buy in" from various sectors and stakeholders.

References

Afrin, S., Chowdhury, F. J., & Rahman, M. M. (2021). COVID-19 pandemic: Rethinking strategies for resilient urban design, perceptions, and planning. *Frontiers in Sustainable Cities, 32*. https://doi.org/10.3389/FRSC.2021.668263

Ajnum, Z. & Nagabhatla, N. 2021. *Reconsidering water-health-migration nexus amid COVID-19: A 5-point strategic framework* [online]. [Cited 28 December 2021]. https://environmentalmigration.iom.int/blogs/reconsidering-water-health-migration-nexus-amid-covid-19-5-point-strategic-framework

Alabaster, G., & Nagabhatla, N. (2020). Human settlements. In *United Nations world water development report 2020: Water and climate change* (pp. 110–117). UNESCO. Also available at https://www.unwater.org/publications/world-water-development-report-2020/

Avellán, T., Nagabhatla, N., Jalan, I., & Liao, D. (2022). Integrating circularity to achieve sustainability: Examples of various wastewater treatment systems. In A. Stefanakis & I. Nikolaou (Eds.), *Circular economy and sustainability* (1st ed., pp. 21–35). Elsevier. https://doi.org/10.1016/B978-0-12-821664-4.00002-9

Bellina, L. (2016). Feeding cities sustainably: The contribution of a 'zero-foodwaste-city' to sustainable development goal 2, 'zero hunger'. In I. A. S. Olsson, S. M. Araújo, & M. F. Vieira

(Eds.), *Food futures: Ethics, science and culture* (pp. 113–118). Paper presented at, 28 September 2016, Wageningen. https://doi.org/10.3920/978-90-8686-834-6_16

Bhattacharyya, S. (2021). Revisiting East Kolkata wetlands-management options with holistic approach. In *Special publication of The Geological Survey of India* (pp. 129–138). Paper presented at, 2021, Kolkata. Also available at https://www.researchgate.net/publication/354992201_Revisiting_East_Kolkata_Wetlands_-Management_Options_with_Holistic_Approach

Boonmahathanakorn, S. (2020). *Rubbish dump turned lush urban farm | International Institute for Environment and Development* [online]. [Cited 13 December 2021]. https://www.iied.org/rubbish-dump-turned-lush-urban-farm

Buchholz, K. (2020). *How has the world's urban population changed?* [online]. [Cited 28 December 2021]. https://www.weforum.org/agenda/2020/11/global-continent-urban-population-urbanisation-percent/

Chandran, R. (2021). Thai landfill turned into urban farm to feed poor during pandemic. 7 January 2021. Also available at https://www.reuters.com/article/us-thailand-landrights-environment-trfn-idUSKBN29C0VH.

Cilliers, E. J., Lategan, L., Cilliers, S. S., & Stander, K. (2020). Reflecting on the potential and limitations of urban agriculture as an urban greening tool in South Africa. *Frontiers in Sustainable Cities, 0*, 43. https://doi.org/10.3389/FRSC.2020.00043

Clos, J. (2015). From COP21 to the new urban agenda. United Nations Chronicle. Also. available at https://www.un.org/en/chronicle/article/cop21-new-urban-agenda.

Dzebo, A., Janetschek, H., Brandi, C., & Iacobuta, G. (2019). Connections between the Paris agreement and the 2030 agenda: The case for policy coherence. SEI Working Paper. Stockholm. Also available at https://cdn.sei.org/wp-content/uploads/2019/08/connections-between-the-paris-agreement-and-the-2030-agenda.pdf.

Eliazer Nelson, A. R. L., Ravichandran, K., & Antony, U. (2019). The impact of the green revolution on indigenous crops of India. *Journal of Ethnic Foods, 6*(1), 1–10. https://doi.org/10.1186/S42779-019-0011-9/TABLES/2

Eltinay, N., & Harvey, M. (2019). Building urban resilience in the Arab Region: Implementing the Sendai framework for disaster risk reduction 2015–2030 at the local level. Contributing paper to GAR 2019. Available at https://www.preventionweb.net/publication/building-urban-resilience-arab-region-implementing-sendai-framework-disaster-risk.

European Bank. (2018). *Belgrade solid waste PPP* [online]. [Cited 13 December 2021]. https://www.ebrd.com/work-with-us/projects/esia/belgrade-solid-waste-ppp.html

FAO. (2015). *The impact of natural hazards and disasters on agriculture, food security and nutrition.* Available at https://www.fao.org/3/i5128e/i5128e.pdf

FAO. (2021a). Towards sustainable and resilient agriculture in Latin America and the Caribbean. Analysis of seven successful transformation pathways. In: *Towards sustainable and resilient agriculture in Latin America and the Caribbean* [online]. [Cited 13 December 2021]. https://www.fao.org/americas/priorities/sustainable-and-resilient-agriculture/panorama-2021/en/

FAO. (2021b). *The state of the world's land and water resources for food and agriculture—Systems at breaking point. Synthesis report 2021.* https://doi.org/10.4060/cb7654en

FAO. (2021c). *The impact of disasters and crises on agriculture and food security: 2021.* https://doi.org/10.4060/cb3673en

Finlayson, M. C., Bartlett, M., Davidson, N., & McInnes, R. (2013). The Ramsar convention and urban wetlands: An opportunity for wetland education and training. In S. Paul (Ed.), *Workbook for managing urban wetlands in Australia* (1st ed.). Sydney Olympic Park Authority. Also available at https://www.sopa.nsw.gov.au/Resource-Centre/WET-eBook-Workbook-for-Managing-Urban-Wetlands-in-Australia

Gondhalekar, D., & Erlbeck, R. (2021). Application of the nexus approach as an integrated urban planning framework: From theory to practice and back again. *Frontiers in Sustainable Cities, 0*, 143. https://doi.org/10.3389/FRSC.2021.751682

Hoover, D. (2017). Report: What, where, and how much food is wasted in cities. Retrieved from Natural Resources Defence Council. Also available at https://www.nrdc.org/experts/darby-hoover/report-what-where-and-how-much-food-wasted-cities.

IIASA. (2018). Tansformations to achieve the sustainable development goals. Report prepared by The World in 2050 initiative. https://doi.org/10.22022/TNT/07-2018.15347

IIED. (2016). The Sendai framework for disaster risk reduction: A critical analysis of the role of cities and local governments in the global agendas. Also available at https://www.citiesalliance.org/sites/default/files/sendai-iied.pdf

Ilieva, R. T. (2016). *Urban food planning : Seeds of transition in the global north* (1st ed.). Routledge.

INE. 2007. Mexico's third national communication to the United Nations Framework Convention on Climate Change. Also available at https://unfccc.int/sites/default/files/resource/MEXICO%20Third%20National%20Communication_English%20version.pdf?download.

INECC & SEMARNAT. (2015). First Biennial update report to the United Nations Framework Convention on Climate Change. Also available at https://unfccc.int/sites/default/files/resource/Executive%20Summary_1.pdf?download.

IPCC. (2018). Summary for policymakers. In V. Masson-Delmotte, P. Zhai, H.-O. Pörtner, D. Roberts, J. Skea, P. R. Shukla, A. Pirani, et al. (Eds.), *Global Warming of 1.5°C. An IPCC Special Report on the impacts of global warming of 1.5°C above pre-industrial levels and related global greenhouse gas emission pathways, in the context of strengthening the global response to the threat of climate change….* Also available at https://www.ipcc.ch/site/assets/uploads/sites/2/2019/05/SR15_SPM_version_report_LR.pdf

Jain, S., Newman, D., Cepeda-Márquez, R. & Zeller, K. 2018. Global food waste management: An implementation guide for cities. Also available at https://www.worldbiogasassociation.org/wp-content/uploads/2018/05/Global-Food-Waste-Management-Full-report-pdf.pdf

John, D. A., & Babu, G. R. (2021). Lessons from the aftermaths of green revolution on food system and health. *Frontiers in Sustainable Food Systems, 5*, 644559. https://doi.org/10.3389/FSUFS.2021.644559/FULL

Kalapurakal, R. (2018). *Cities 2030: Implementing the new urban agenda* [online]. [Cited 30 December 2021]. https://www.undp.org/blog/cities-2030-implementing-new-urban-agenda

Mayhew, M. (2020). One of Europe's largest landfills gets a makeover, May 2020. Also available at https://www.ifc.org/wps/wcm/connect/news_ext_content/ifc_external_corporate_site/news+and+events/news/cm-stories/serbia-waste-to-energy.

Mouratidis, K., & Yiannakou, A. (2021). COVID-19 and urban planning: Built environment, health, and well-being in Greek cities before and during the pandemic. *Cities, 103491*, 103491. https://doi.org/10.1016/J.CITIES.2021.103491

MUFPP. (2020). *The Milan pact—Milan urban food policy pact* [online]. [Cited 1 January 2022]. https://www.milanurbanfoodpolicypact.org/the-milan-pact/

Nagabhatla, N., & Metcalfe, C. D. (Eds.). (2018). *Multifunctional wetlands* (1st ed.). Environmental Contamination Remediation and Management. Cham, Springer International Publishing. https://doi.org/10.1007/978-3-319-67416-2

Nesbit, M., Paquel, K., & Illes, A. (2017). Research for REGI Committee—Cohesion policy and Paris Agreement Targets. https://doi.org/10.2861/09402

Pingali, P. L. (2012). Green revolution: Impacts, limits, and the path ahead. *Proceedings of the National Academy of Sciences, 109*(31), 12302–12308. https://doi.org/10.1073/PNAS.0912953109

Ramsar. (2016). *An Introduction to the Ramsar Convention on Wetlands* (7th ed.). Gland, Ramsar Convention Secretariat. Also available at https://www.ramsar.org/sites/default/files/documents/library/handbook1_5ed_introductiontoconvention_final_e.pdf

Rekow, L. (2020). Integrating food culture with socio-environmental recovery: Case study perspectives from the global south. *Humanities, 9*(4), 134. https://doi.org/10.3390/H9040134

RUAF. (2020). Cities, climate change and the food system of the majority. Retrieved from RUAF Blogs. Also available at https://ruaf.org/news/cities-climate-change-and-the-food-system-of-the-majority/.

Sadashivam, T., & Tabassu, S. (2016). Trends of urbanization in India: Issues and challenges in the 21st Century. *International Journal of Information Research and Review, 03*(05), 2375–2384. Also available at https://www.researchgate.net/publication/332103931_TRENDS_OF_URBANIZATION_IN_INDIA_ISSUES_AND_CHALLENGES_IN_THE_21_ST_CENTURY

Sauvé, S., Lamontagne, S., Dupras, J., & Stahel, W. (2021). Circular economy of water: Tackling quantity, quality and footprint of water. *Environmental Development, 39*, 100651. https://doi.org/10.1016/J.ENVDEV.2021.100651

Singh, A. K. (2009). Urbanisation, Urban Growth and Urban Management in India: Shifting Paradigm. In R. S. Singh (Ed.), *Indian geography in the 21st century: The young geographers agenda*. Cambridge Scholars Publications. Also available at https://www.researchgate.net/publication/319988171_Urbanisation_Urban_Growth_and_Urban_Management_in_India_Shifting_Paradigm

Singh, R. B. (2000). Environmental consequences of agricultural development: A case study from the green revolution state of Haryana, India. *Ecosystems and Environment, 82*.

Steenkamp, J., Cilliers, E. J., Cilliers, S. S., & Lategan, L. (2021). Food for thought: Addressing urban food security risks through urban agriculture. *Sustainability, 13*(3), 1267. https://doi.org/10.3390/SU13031267

Tollin, N., Hamhaber, J., Grafakos, S., Lwasa, S. & Morato, J. 2017. Sustainable urbanization in the Paris Agreement: Comparative review for urban content in the Nationally Determined Contributions (NDCs). . Also available at https://unhabitat.org/sites/default/files/download-manager-files/Paris%20Agreement_25_Jan-2019_for%20Email-Online.pdf

UN. (2015). *Paris Agreement*. United Nations, [Cited 29 December 2021]. https://unfccc.int/sites/default/files/english_paris_agreement.pdf.

UN. (2017). New Urban Agenda. (also available at https://uploads.habitat3.org/hb3/NUA-English.pdf).

UN. (2019). World urbanization prospects: The 2018 revision. . (also available at https://population.un.org/wup/Publications/Files/WUP2018-Report.pdf).

UNECE. (2020). People-smart sustainable cities. . Also available at https://unece.org/sites/default/files/2021-01/SSC%20nexus_web_opt_ENG_0.pdf.

UNECE. (2021). *"People-smart" and "cities-based" approach to sustainable development is key, urges UNECE report* [online]. [Cited 28 December 2021]. https://unece.org/circular-economy/press/people-smart-and-cities-based-approach-sustainable-development-key-urges

UNEP. (2020). *"Wetlands and biodiversity" is the theme for world wetlands day 2020* [online]. [Cited 28 December 2021]. https://www.unep.org/news-and-stories/story/wetlands-and-biodiversity-theme-world-wetlands-day-2020

UNEP. (2021). Food waste index report 2021. . Also available at https://www.unep.org/resources/report/unep-food-waste-index-report-2021.

UNEP. undated. *Worldwide food waste* [online]. [Cited 28 December 2021]. https://www.unep.org/thinkeatsave/get-informed/worldwide-food-waste

UNFCCC. (2016). Report of the conference of the parties on its twenty-first session, held in Paris from 30 November to 11 December 2015. Also. available at https://unfccc.int/resource/docs/2015/cop21/eng/10a01.pdf.

Vasovic, A. (2021). Serbia closes largest unmanaged landfill in Europe -report, 26 August 2021. Also available at https://www.reuters.com/world/europe/serbia-closes-largest-unmanaged-landfill-europe-report-2021-08-26/.

Vicziany, M., Chattopadhyay, D., & Bhattacharyya, S. (2017). Food from sewage: Fish from the East Kolkata wetlands and the limits of traditional knowledge. *Journal of South Asian Studies, 40*(3), 619–644. https://doi.org/10.1080/00856401.2017.1341038

World Bank. (2019). *Improving energy for agricultural competitiveness in Mexico* [online]. [Cited 13 December 2021]. https://www.worldbank.org/en/results/2019/05/10/improving-energy-for-agricultural-competitiveness-in-mexico

Xu, H., & Tuts, R. (2021). *Five years on, taking stock of the new urban agenda* [online]. [Cited 31 December 2021]. https://www.undp.org/blog/five-years-taking-stock-new-urban-agenda

Chapter 15
Summary, Recommendations, and Roadmap for Global Food Security

Amani Alfarra, Atif A. Kubursi, and Nathaniel K. Newlands

Abstract We synthesize the critical insights and recommendations emerging from this comprehensive volume, which addresses the multifaceted challenges to global food security in an increasingly crisis-prone world. We highlight the interconnected impacts of major global disruptions, such as the COVID-19 pandemic, the Ukraine conflict, and escalating climate change, on food systems, biodiversity, and urban sustainability. We underscore the urgent need for innovative, inclusive, and resilient approaches to agri-food systems that align with the Sustainable Development Goals (SDGs). Our key recommendations include advancing sustainable practices like Nature-based Solutions (NbS), circular agriculture, and digital transformation to strengthen urban and peri-urban food systems. Emphasis is placed on fostering local food production, reducing waste, enhancing resilience, and promoting equity in food access and nutrition. We advocate for multi-stakeholder collaboration and policy interventions to address systemic vulnerabilities, drive green transitions, and secure food supply chains amid increasing uncertainties. The integration of these strategies provides a roadmap for achieving sustainable food security for both urban and rural communities worldwide.

A. Alfarra (✉)
Food and Agriculture Organization of the United Nations, Land and Water (NSL) Division, Rome, Italy

Walter G. Booth School of Engineering Practice and Technology, Hamilton, ON, Canada
e-mail: Amani.Alfarra@fao.org

A. A. Kubursi
Department of Economics Kenneth Taylor Hall (KTH), McMaster University, Hamilton, ON, Canada

N. K. Newlands
Agriculture and Agri-Food Canada, Summerland Research and Development Centre, Summerland, BC, Canada

Department of Geography, University of Victoria, Victoria, BC, Canada
e-mail: nathaniel.newlands@agr.gc.ca

A. A. Kubursi et al. (eds.), *Urban Food Security in a Crisis Prone World*,
https://doi.org/10.1007/978-3-031-89440-4_15

Keywords Agriculture · Crisis · Food security · Resilience · Sustainable Development Goals · Urban agriculture

Synthesis

The overarching theme of the book is the focus on the crucial impacts of the crisis-prone world on urban food security. This theme became the central idea of the book and it tied together the many diverse contributions into a consistent and rich mélange of ideas, diagnoses, and prescriptions. It is to be expected that a compendium that brings together noted scholars and experts in disparate fields including Economists, Engineers, Physical Scientists, Agronomists, Social Scientists, and Environmentalists would bring multiple perspectives based on the many disciplines and expertise knowledge of the contributors. In the next sections, a summary of the salient arguments and recommendations of each chapter is presented. The aim is to draw and highlight the important messages and recommendations tendered in each chapter.

Chapter 1. Crisis-Prone World, the Urban Food-Water Nexus: An Introduction

The pandemic presented new challenges of its own to the world but particularly to the Global South. It affected directly the availability, accessibility, and affordability of food. Strengthening structurally the accessibility and stability of food systems and curbing food price escalation formed the key areas for policymakers to address the pandemic's impact on lives and livelihoods in general and in the urban centers in particular.

The conflict in Ukraine, coming on the heels of the pandemic exacerbated its negative impacts. Disruptions of the food supply chains, especially food products from Russia and Ukraine,[1] compromised the structural transformations of food production worldwide and provided strong incentives for home-based and local food production. Several countries opted to reduce waste, build storage capacities, and tried to connect local consumers with local producers with the view to reduce dependence on imports. This strategy of ensuring more secure sources of food and tying cities to their surroundings as sources of reliable food supplies also created many income generation opportunities.

[1] These two countries generally contributed over 25% of world trade in wheat and 70% of corn and for some countries like Egypt, Russia, and Ukraine supplied over 50% of their food imports of these commodities.

The pandemic provoked the reevaluation and restructuring of food systems toward increased dependence on local production based on Green initiatives such as Nature-based Solutions (NbS). The latter promoted the redesign of food systems and agriculture production with the aim to minimize their ecological footprint and to improve food security. NbS is favored as it offers potential opportunities to tackle food security while ensuring ecological quality and enhancement of green infrastructure, primarily in urban areas.

One of the valuable lessons learned from the pandemic is the realization that food forms a vital sector in the global economy, and that the food system can be fundamentally redesigned and restructured for multiple benefits. Reconnecting communities with local food production can actively tackle simultaneously climate change and biodiversity loss, promote human health, reduce food insecurity, and improve the resiliency of communities.

Reducing edible food waste has the potential to reduce CO_2 emissions coupled with substantial savings. Food product developers and others responsible for designing food systems could transform typically food by-products into food products.

More than 40% of the world's irrigated cropland is located in peri-urban areas, yet the food produced on this land is exported while similar products are imported into neighboring cities. By reconnecting cities with surrounding peri-urban areas, a circular economy for food can be built resulting in stronger resilience in the food system, improves food security, while saving money, reducing emissions, and creating jobs.

Sourcing of locally grown food using regenerative practices (where appropriate) could generate large financial benefits, reduce food waste, and promote the conversion of remaining organic waste into high quality compost, which subsequently yields health benefits and reduce health costs.

Chapter 2. Redesigning and Transforming Food Systems to Be More Resilient to Disruption and Disaster: Lessons Learned from COVID

A sustainable food system provides food security and nutrition for all in such a way that the economic, social, and environmental systems involved in generating food security and nutrition for future generations are not compromised. Any disruption in the food system will have ripple effects that stretch wide and affect negatively education, health, the economy, human rights, peace, and security.

Structural changes in the food system (e.g., diversified products with longer shelf life) and the use of digital agricultural technologies, such as artificial intelligence (AI) and block-chain systems represent some of the broader post-COVID transformations that could be undertaken to adapt and build resilience to the pandemic's negative impacts.

Key lessons learned from COVID that apply also to addressing climate change include the adoption of digital technology in communities hoping to stay operational and resilient, and the value of cooperation, maintaining transparency and trust among stakeholders across all levels of the food system.

There is an urgent need to transition from the current dominant food systems that were designed for mass production of standardized foods without integrating social, economic, and environmental sustainability, to a more sustainability-themed system with a strong focus on efficiency, substitution, and integration of conservation agriculture.

Accelerating the transition to a more globally sustainable food system requires increased adoption of and innovation in digital technologies by the public and private sectors, sustainable (integrated) watershed management, and policy-based incentives with robust public policy leadership alongside consumers who culturally value and can afford healthy, sustainable diets.

Digital agriculture applied to diverse food system types can drive the transformation, redesign of agriculture, and support the efficient management of natural resources and better environments. Policy makers, citizens, and relevant stakeholders should promote, enhance, and expand a digital inclusive agenda that encompasses the availability of data-driven and evidence-based agriculture.

Digital technologies in agriculture are becoming more used but there is still the need for better matching to farmers' needs. In addition, there is room for better integration of local expert knowledge at the farm level. At the scientific community scale, there is also a need for broader data sharing to enable improved data validation and calibration for better predictive analytics.

Chapter 3. Environmental Resilience and Agricultural Best Practices

With the expected rise in global temperature, the likelihood of the occurrence of algae blooms in water bodies around the world has significantly increased. This poses serious problems for the production of food, particularly near urban centers close to water. Since the mid-1990s, a resurgence of planktonic cyanobacterial harmful algae blooms has been noted in Lake Erie, Ontario, Canada. Increased loadings of phosphorus in a highly reactive form (dissolved reactive phosphorus) are thought to be in large part responsible for the return of these blooms, not seen since the 1970s. Increased summertime temperatures are also a factor.

If people drink water contaminated by microcystins (the toxin produced by *Microcystis aeruginosa*), symptoms of exposure include nausea, vomiting, and, in rare cases, acute liver failure. Minor skin irritation can also occur with contact. Reported health effects from cyanobacteria in humans are, however, uncommon in the United States and Canada. The effects of toxic algae on fish are mostly due to the hypoxia induced by decaying algae rather than from the algae itself.

As with hazardous blooms, increased loadings of dissolved reactive phosphorus—mainly from the runoff of fertilizer from farmland—are a major cause of the growth of nearshore nuisance blooms. Other factors are increased summertime temperatures and, in the case of *Cladophora*, the presence of dreissenid (zebra and quagga) mussels, which filter water (allowing greater light penetration at depths where the algae form) and provide a substrate for algal growth.

To reduce the severity of HNABs, the governments of Canada and the United States have committed to reducing phosphorus loads to Lake Erie by 40% from 2008 levels and coordinate action through a binational framework developed under the Great Lakes Water Quality Agreement.

The study demonstrated that rigorous evaluation of the economic impacts of agriculturally related freshwater degradation is possible and that the size of the impacts—at least in the context of Lake Erie—is sufficient to warrant the attention of government policymakers. At the same time, the study revealed a variety of areas where both the physical and economic understanding of HNABs could be improved.

Chapter 4. Peri-Urban Agriculture and Food Security in MENA Countries

Among the major challenges facing the peoples of the MENA (Middle East and North Africa) region include the high dependency on agricultural and agri-food imports and the slow increase of agricultural production compared to food demand due to agronomic and climate constraints, in addition to the limited impact of agricultural policies.

The MENA region is characterized by the strong growth of its urban population, a high proportion of irrigated agriculture due to low precipitation levels, low levels of irrigation efficiency, and the increased competition among water users. These have posed a major concern for agricultural systems in the region.

Developing Urban Agriculture (UA) and Peri-Urban Agriculture (PUA) in the MENA (Middle East and North Africa) region requires increased recognition by policymakers and stakeholders of its potential benefits and challenges and the need for more scientific research to evaluate these potential benefits than hitherto has been the case.

Urban Agriculture (UA) has the potential to complement conventional agriculture and contribute to food and nutritional security without burdening the food system impacts in the MENA region. This is based on the analytical results from foresight studies, literature review, and empirical analysis of global data.

Chapter 5. Self-Provisioning and Urban Agriculture

The urbanization of the Canadian population has been a significant force in reducing self-provisioning. The COVID-19 pandemic triggered increased interest in gardening and promoted changes in food production practices to address individual and community vulnerability to disruption of food supply chains.

There are still a number of impediments to increased self-provisioning, which include a lack of the main production inputs, stringent food safety rules, and negative cultural perceptions. A few benefits can be expected from self-provisioning and these include better nutrition, better health, economic savings, and greater community cohesion and cultural expression.

Provision of main production inputs, collaboration among main private and public stakeholders, and availability of volunteers and experts for technical support constitutes some of the main necessary ingredients for successful municipal gardening programs as exemplified in Montreal, Canada.

Because beekeeping and chicken rearing often happen in backyards, it is challenging to develop a clear picture of levels. The 2015 survey of 81 Canadian mid-sized and large cities found that almost 90% permitted urban beekeeping, almost one-fourth permitted urban chicken rearing, and about 7% had urban aquaculture operations present, presumably commercial Toronto has about 500 beehives in about 200 yards (Secord, 2019). Some Montreal boroughs, Victoria, Vancouver, Kingston, Kitchener, Niagara Falls, Toronto (pilots), Edmonton, and Whitehorse are some of the cities that permit chickens.

Cities produce significant amounts of food (backyards, municipal properties, rights of way, and public trails) that often go unharvested, particularly fruit and nuts, and other kinds of edible landscapes. This has spurred on urban foraging movements; sometimes individual, increasingly through nongovernmental organizations and social purpose enterprises, such as LifeCycles in Victoria.

Chapter 6. Economic Commercial Vertical Urban Farming

The vertical farming concept can optimize food production to address imminent global challenges such as population growth, urbanization, soil erosion, decreased agricultural yields, reduced precipitation, and increased greenhouse gas emissions. Vertical farming technology is potentially the only feasible solution to address food security, supply chain, and food safety issues at once.

Vertical farming technology is generally undertaken in two stages: the first is the production of varied types of leafy green vegetables, which is currently successfully practiced. The second stage is the production of common vegetables. Both stages have room for further development and automation to lead to the optimization of all the process parameters.

The fruiting process of some vegetables requires considerable amounts of energy; hence the limited production and ongoing research are focused on how to venture into their production. It is expected to take about two decades for it to be feasible, but challenges posed by the pandemic could shorten this time span.

Space and resource limitations will spur farmers and relevant stakeholders to explore innovative solutions to address production challenges. Vertical farming technology is expected to be one of the main solutions available and developments in the fields of renewable energy, artificial lighting, robotics, and AI will contribute to that end.

Chapter 7. The Role of Livestock Production in Urban Farm

Using a number of case studies, it was confirmed that urban livestock production systems can stabilize smaller communities economically, boost food security, and provide environmental benefits. These systems have proven to be resilient to a wide variety of shocks, hence the need for structured investment in them to increase food resilience in urban areas.

Successful urban agriculture requires robust institutional support effort and access to veterinary services, financing from the public and private sectors, adoption of sustainable best practices, and sound coordinated urban development planning initiatives.

More research is needed to build up the resilience levels of urban livestock production systems. Some small-scale urban and peri-urban livestock production systems have proven to be resilient to shocks resulting from the COVID-19 pandemic. This is why it would be best to capitalize on these systems to contribute to efficient and effective urban food systems for the future.

If urban livestock production is to be sustainable and contribute to the food and environmental systems of the world's growing urban areas, it is clear that more effort needs to be made to translate traditional livestock production methods and best practices into a more complex urban setting. Conventional livestock production systems are often based on spatially spread out systems with livestock production relegated to low population density rural areas, relying on improved transportation and market infrastructure to move livestock products to urban consumer centers. These systems are responsible for producing affordable livestock products for urban consumers worldwide and it is unlikely that urban livestock production systems will be able to compete with conventional systems that can capitalize on economies of scale and have a long history of research and investment in improved practices and efficiencies.

However, current research into conventional livestock production has also identified important negative environmental externalities associated with this approach, such as excess accumulation of animal waste, runoff into surrounding areas and waterways, and reduced nutrient cycling to intensive cropping systems due to transport costs. Other research has highlighted potential negative animal welfare consequences of some intensive conventional livestock production practices that are increasingly questioned by urban consumers. And finally, dependence on long and growing conventional livestock value chains also has risks, as can be seen in the recent experience of the COVID-19 pandemic, which disrupted perishable livestock value chains worldwide, with resultant threats to farmers and urban consumers alike.

Fortunately, research is developing into different ways to minimize negative externalities and risks in conventional livestock production that may also have potential in urban areas. For example, integrated crop-livestock systems (ICLS) minimize environmental costs of livestock production by allowing animals to eat locally available food and the use of fertilize crops in situ, rather than outsourcing both, often from synthetic sources requiring energy to manufacture and transport.

Chapter 8. Organizing Urban Farms in Post-COVID Recovery

Despite its many adverse effects, the COVID-19 pandemic mobility restrictions strengthened joint ventures of urban farmers to buy inputs, produce, and transport and sell their produce collectively after negotiations with the relevant authorities. Such effects inform the need for improved urban farmers' cooperation.

The fragility of urban areas to food security during the pandemic lockdowns highlights the need to bring urban agriculture to feed communities in these areas, diversify food systems, and increase resilience.

Organization of urban farmers and their commitment to well-structured cooperatives can improve their negotiation power with the relevant authorities to improve their capacity to buy inputs and access better market opportunities.

Nongovernmental organizations can support urban farmers' organizations through capacity building on technical aspects, financial support and subsidies, and conflict management. Governmental support should accord an enabling policy environment that encourages investments in agriculture.

By examining the principles of the International Cooperatives Alliance (ICA),[2] it becomes clear that more efforts should be put into rules that define common resources' sharing, control of the management board by members, and economic participation of members and revenues retribution to members. These efforts should start with the urban farmers themselves as a commitment to succeed collectively. Then, specialized nongovernmental organizations can enter the fray and support the process by building or strengthening the capacity of groups of farmers in group management. Two key issues that would need special attention during these capacity-building programs are financial management and conflict management, as also recommended by some previous studies that analyzed important factors affecting the proper functioning of smallholder cooperatives in Benin and South Africa.

Moreover, the support of agricultural and economic policies is necessary by establishing a guarantee fund, which may encourage financial institutions to take more risks in agriculture financing, thereby enticing urban farmers to improve their cooperation. This policy support is also identified by previous studies as important to create an enabling policy environment for urban farmers. Finally, lessons learnt from this chapter would be useful to inform a strategy for cooperatives development in a post-COVID-19 economic recovery.

[2] ICA Principles: https://ica.coop/en/cooperatives/cooperative-identity

Chapter 9. Selecting Allotment Gardens in Urban and Peri-Urban Areas

The downside of the current rapid urbanization is well known; cities grow faster than their minimum service levels can cope with. This is particularly manifested in urban food insecurity, particularly in the lower wealth segments of the Global South countries. The impoverished urbanized population, the world over, had to face even greater challenges under the COVID-19 mobility restrictions that disrupted food supply chains while those who lose their jobs in the informally organized sector do not qualify for income compensation.

Urban agriculture seems a viable answer; it can be practiced at safe and infection-preventive distances, produces healthy and nutritious food, earns some extra income from surplus production, and can be located away from infection-prone mass concentrations. Political support to implement allotment gardens is largely absent and continues to be sustained by information gaps on the allocation and suitability of potential sites.

This chapter addressed this knowledge gap by presenting the fundamental building blocks for a site allocation tool that optimizes the spatial planning of allotment gardens in urban and peri-urban areas while accounting for biophysical (soil, water) and socioeconomic (distance to markets and safe entrance for women) conditions as well as COVID-19 safety measures and will provide better living conditions.

Planning tools can support city administrators to plan urban allotment gardens near cities and peri-urban areas. The design and development of the allotment system should take into account the independent and dependent variables, suitability assessment of the proposed sites, and a provision for policy interpretation of outcomes, in addition to a strong political will with a clear supervision mandate.

Chapter 10. Tracking Ecological Overshoot: The Food System's Ten "Impossible Imperatives"

The food system, and particularly farms, will have to produce all the food they need while rapidly phasing out fossil fuels or its derivatives from farm to fork. Humanity, if it wants to comply with the Paris Agreement and avoid the worst of climate change will have to get out of fossil fuel use well before 2050. Even if humanity does not want to, eventually it will be forced to quit its fossil fuel dependence, and will face even more unpredictable climate extremes, if it were to phase out those fuels slowly.

As food is one of the basic human needs, the global food system also faces an imperative to contribute to food equity, challenged by a large number of people with no access to healthy diets and food insecurity. There is an increasing demand for food from a growing population that is increasingly urbanized. The demand is getting amplified by shifting diets: many people seem to want ever more protein in their diet, especially animal-based protein—even though they may not have a protein shortage in their diet. Typically, protein calories take far more bio-capacity than general food calories (such as starches), even if the protein is from plants rather than animals. Some of the increased food demand can be provided by cutting food loss and waste.

Currently, estimates suggest that approximately one-third of food produced is wasted. Also, the increase in demand, which may well exceed the growth rates of populations, may be so high that there will be a need for higher yields. Some may argue though that increases in population size, even globally, may not need to occur.

Farms will have to operate in ways that do not erode or contaminate their soils and waterways, now and in the future. Given our fragile planet, humanity can no longer afford to degrade or lose soil, overexploit and deplete surface and groundwater, assault biodiversity, or apply substances emitting persistent pollutants. There may even be calls for farms to become regenerative, agro-ecological, or even restorative.

Additionally, to live up to IPCC scenarios, agricultural land must become a sink of GHG, not a source as it is now, particularly post-2050. This means that farms will have to change practices that emit additional greenhouse gases, particularly methane.

We can no longer expect and depend on the usual, regular weather patterns, the food system must become climate resilient. Extremes will be more common, overlaid, most likely, with an overarching warming trend (at least as a global average). The average climate may be heating up more, or less, in your own region, with increased seasonal variance. This may affect the evapotranspiration conditions of what you grow. Therefore, it may require significant adaptation efforts, possibly growing different kinds of crops, shifting location of production locations to the urban areas and adjusting the contours of your territories to cope with changing water availability.

Chapter 11. Peri-Urban Agriculture and Food Security in MENA Countries

COVID-19 escalated from a public health emergency status to be declared by the WHO on the 11 March 2020 as a global pandemic. The pandemic has been described as "an extraordinary situation" where it represents a crisis of a unique nature that has affected all countries across the globe regardless of their economic status (ITC, 2021). Analogously within each country, most of the segments and actors in the economy were negatively affected, though with different intensities. COVID-19 crisis is a multidimensional crisis, where at the outset it is a health crisis, but at its core it is an economic as well as food security (FS) crisis, hence making the world featuring a triple shock.

The most substantial increase in the number of people suffering acute hunger in 2020 is likely to be concentrated in the least developed countries (LDCs) of sub-Saharan Africa, and developing economies of Middle East North Africa (MENA) and Latin America. Some forecasts suggested an addition of 83–132 million people to the ranks of the undernourished in 2020 due to COVID-19.

Some rectifying measures were taken including expanding cash transfers to the needy and consumer subsidies. Furthermore, a few additional proactive measures were implemented that included increasing the storage capacity. These measures helped to contain some of the pandemic negative impacts.

COVID-19 adversely affected the status and operating performance of the food systems in most of the countries in the MENA region. A few suffered reduced food consumption, availability problems, and labor shortages. However, agricultural production of some crops such as wheat experienced an increase due to higher government procurement prices. Still a few measures are needed to tackle food accessibility and affordability.

Digitization has proven to be effective in shielding many sectors against the worst effects of the COVID-19 pandemic. Intensifying the usage of digitalization within food supply systems can be further exploited in the MENA region in response to the pandemic.

Empirical assessments showed that in the MENA region, COVID, whether expressed by its intensity, or by the degree of country preparedness, is significantly linked to the countries differences in their food system levels. Transparent dissemination of COVID-related data, e.g., cases, food prices, etc., will strengthen management of food markets and lessen unfavorable outcomes in food systems.

MENA countries should make use of this crisis to tackle the structural imbalances affecting accessibility and the utilization pillars of the food system. COVID has brought time for MENA countries in general and for GCC countries in particular by allowing them to shift their major focus from food availability to food accessibility and utilization. The former efforts in MENA countries have left those two dimensions of the food system lagging behind. Measures to tackle the affordability and accessibility pillars have been general and not food focused. In this respect, the enhancement of social safety nets as a measure undertaken by all governments might have positive spillover effects on the accessibility pillar, but it remained short on targeting food security per se. Such enhancement should not include civil society charity initiatives as a main resort, if the aim is to at least maintain the level of food sufficiency. This channel will soon dry up if the pessimistic expectations of expanding economic downturn came into effect. Among the policy recommendations that can be implemented in this respect is nudging their population toward a healthy diet and nutrition prescriptions. This might be done through lowering subsidies directed to unhealthy food items and/or providing healthy relatively cheap substitutes. Given the fact that the pandemic came in the right time for consumers to value the importance of consuming healthy food, reforming social safety nets should be tilted to focus on enhancing such kind of awareness. Moreover, MENA countries should devote extra efforts to fix their local supply chains and enhance their integration in global value chains when it comes to food products and inputs, while developing the accompanying risk management systems and overcoming logistical problems.

Chapter 12. Aligning Urban Farming with Green City Aspirations

Peri-urban agriculture can reduce the environmental foot-print associated with shipping produce into a green city. As exemplified by studies in Ontario, Canada, a reduction in food imports could contribute to a significant reduction in CO2 emissions associated with the transportation of these commodities.

Peri-urban farming presents several advantages to entrepreneurs including proximity to consumer markets, low transportation costs coupled with low carbon footprints, access to labor, and potential higher returns compared to traditional farming operations. Peri-urban agriculture can include field production of fruits and vegetables, greenhouse production, and as well vertical farms.

Favorable land-use policies could support the potential of peri-urban agriculture. Greenbelts around a Metropolis, which restrict certain types of land use and/or restrict changes in zoning from agriculture to other uses, are most likely necessary to enable a significant amount of peri-urban farming.

There are significant demands for farmland in the urban fringe where retiring farm operations will, in most cases, want to sell their lands to the highest bidder, which typically includes the best use of the land given the local zoning regulations and land-use restrictions. If there are few zoning restrictions, the highest and best use in such an environment can be based on a developer's use of the land for a housing subdivision or other nonfarm commercial enterprises. This works against the development of peri-urban agriculture in the urban fringe.

This is why Greenbelts around a metropolis, which restrict certain types of land use and/or restrict changes in zoning from agriculture to other uses, are most likely necessary to enable a significant amount of peri-urban farming. In the absence of such greenbelt policies, the highest and best use of land in the urban fringe is typically more urban development, which is at the expense of peri-urban agriculture.

In Ontario, a significant portion of the GH includes the Greenbelt. The Greenbelt surrounds the western shores of Lake Ontario, from Niagara-on-the-Lake, through the GTHA, to parts of Northumberland County. The purpose of the Greenbelt Plan is to permanently protect farmland and support the agricultural system, as well as the natural system and features within it.

Chapter 13. Global Impacts of the Ukraine Invasion: A Resource Perspective

The Ukraine invasion comes with heart-breaking human suffering. It will also impact people far away from the war as global supply chains for food, metals, and energy are disrupted. The price shocks they have caused may well lead to grave consequences, particularly in lowest-income countries, with the potential to amplify conflict and migration pressure. Most of the poor countries in the Global South are heavily indebted and the inflation triggered by the war and the way central banks are dealing with it have raised interest rates to unprecedented levels not experienced in the past 40 years. Many poor countries are not only paying high prices for their food and energy, but also are forced to allocate large parts of their tax revenues to pay for their debts.

Add to this are the already manifested disruptions of climate change in biodiversity loss, freshwater shortage, deforestation, soil degradation, desertification, and food crises; it is becoming increasingly evident that the most limiting physical factor for humanity is not the nonrenewable stocks, but rather the renewables, in particular, biological resources that are massively overused.

The Ukraine invasion adversely affected global supply chains for food, metals, and energy. Cropland dominates Ukraine's exports and the impacts of disruption of the food supply chains resulted in adverse effects on food access, availability, and prices to its main export countries.

For context, Ukraine contains 2.7 global hectares of biocapacity per person, of which, nearly two (1.9 global hectares per person) are cropland. This is about four times the world average. In contrast, the Russian Federation contains 6.7 global hectares of bio-capacity per person, of which only 1 global hectare is cropland demanded by countries. The food portion of the footprint and the bio-capacity follows a similar logic as the entire bio-capacity and footprint.

Even more revealing is the proportion of wheat and cereal consumed worldwide that originates from the Russian Federation and Ukraine. Aside from the source countries themselves, which we would expect to be the top consumers as well, the other top countries on this list are nearby neighboring Georgia and several MENA countries. In Tunisia, the country where the Arab Spring started in 2011, nearly one quarter of its consumed cereals and wheat originated in Ukraine, this proportion is higher in Egypt.

Let us explain how one can interpret statistics like "over 60% of 'wheat and cereals' consumed in Georgia originates from Ukraine and the Russian Federation." Does this mean that Georgia is irrevocably dependent on Ukraine and the Russian Federation? Looking at resource accounting data, as provided here, offers some insights. But it also brings up more relevant questions to consider when analyzing a country's resource dependency and determining associated risks. For instance, do these countries have price-competitive alternatives to Ukrainian crops? Given a lower level of price elasticity of basic food, could this significantly affect market prices, or even outprice a large number of people from accessing basic calories? How will global markets and production systems all over the world adjust in the short and long term? Once we recognize how significant resource security is to provide for robust prosperity, these numbers help us answer some basic questions and formulate even more relevant ones.

Chapter 14. Urban Food Sustainability and Multilateral Environmental Governance Frameworks: Directional and Operational Complexities and Opportunities

Exploring alternative options to improve the production and distribution systems for food and consumable goods remains crucial to achieving food sustainability goals. COVID-19 continues to pose challenges to nations with regard to food systems; all stakeholders should play a collective role in safeguarding sustainable food production.

The nexus approach incorporating energy-food-waste-land as interconnected systems coupled with digital technology that promotes automation and the use of artificial intelligence plays a critical role in designing sustainable urban management strategies.

Expanding urban populations increasingly pose a threat to agricultural land use and can stimulate land-use changes in rural landscapes. If uncontrolled, the urbanization can shift the socioeconomic variables of land and water resources and affect agricultural production.

A food-secure future for urban communities should encompass peri-urban sustainability objectives that have the potential to address environmental and climate change impacts, connect with local and traditional knowledge systems, address context-based norms, integrate aspects of social justice, water, and meet food security targets.

Concluding Remarks

The COVID pandemic alone was sufficient to present a serious challenge to the world economy, globalization, and food systems. The war in Ukraine has presented an additional shock to the world systems that exacerbated the damages generated by the pandemic. The two crises were sufficiently damaging and destructive in their impacts on the global economy, particularly on the economies of the Global South and Europe.

The escalating negative manifestations of climate change are with us now adding cumulative costs as the sea levels rise, unpredictable climate, fires, cyclones, droughts, and an overheated planet. The three crises are coming together when one of which is sufficient to have created major and devastating consequences.

The insights gained from the chapters are helpful to deal with these crises. FAO has advanced a few measures and actions to address some of these global crises. They are summarized below. They may not be sufficient to rectify and deal with all of the challenges and damages; they offer a nuanced approach to deal with them.

- To ensure global food security to meet the demands under uncertainty due to COVID-19 pandemic and increasing population, food production must rise by 60 percent by 2050, hence the need for alternative farming techniques. The hydroponic and vertical farming technologies with closed water loop system are examples that have potential for enormous water and land-use savings, as seen from FAO case studies in the projects in the Gaza Strip and Jordan.

- Reliable Water, Sanitation and Hygiene (WASH) facilities are essential to containing the spread of viruses. The use of nonconventional water (NCW) resources could be beneficial for water and sanitation when made available. Multiple Uses of Water Services (MUS, a characteristic of irrigation systems) presents an opportunity to increase water efficiency, productivity, and reduce water scarcity at different scales and its promotion is particularly timely in the light of the spreading COVID-19. FAO continues to invest in MUS and develop tools and methodologies for auditing and modernizing MUS in irrigation systems with the aim to obtain full recognition in the policy arena of the multi-benefits of MUS.
- Lockdown measures drove a major shift toward the "circular economy" approach. A "circular agriculture economy" proposes a viable model for the current linear economy "take-make-waste" approach by minimizing the amount of external inputs for agricultural production, closing nutrient loops and reducing negative impacts to the environment. FAO promotes the opportunities presented by "circular agriculture economy" including water reuse, recycling of irrigation water, optimized management of inputs in a field according to actual crop needs (i.e., precision agriculture), and the use of bio-fertilizers and bioenergy.
- FAO supports national governments and local city planning in shifting toward greener and regenerative cities' systems through facilitating digital transformation by providing geospatial technologies to identify hotspots of vulnerability, and supporting nature-based solutions as well as urban and peri-urban horticulture and the adoption of sustainable land, soil, and water management practices. Practical examples on taking action by FAO and its partners for resilient and regenerative cities include the "Enhancing Urban Horticulture Production in Bangladesh" and "Greening infrastructure to address water security in Lima, Peru" projects, which contributed to generating local income opportunities and increased food security.
- FAOs World Agriculture Watch (WAW) offers a comprehensive approach and tools to collect data on family farms, which allows identifying needs and targeting investments in three stages. The first stage is generating evidence on family farms that can help develop country response plans and assess family farmers' needs. The next stage is using the evidence to direct investments based on the collected data to redirect the necessary funds to the benefit of family farmers. The last stage establishes a monitoring system to assess investment impacts. Pilot case studies have been carried out by FAO in El Salvador and Mali.
- FAO Land and Water is playing a key role by assessing how the COVID-19 pandemic is affecting the food and agriculture sector with an array of tools for supporting policy analyses and for assessing impacts of COVID-19 on food and agriculture, value chains, food prices, and food security across the globe. Examples of databases and tools are Global Agro-Ecological Zoning (GAEZ) tool, Water Productivity through Open Access of Remotely Sensed Information (WaPOR), and Land Resources Information and Management Systems (LRIMS).

With the emerging global challenges, there is a need for increased concerted efforts and collaboration to bring together all stakeholders: from governments, private sector, NGOs, and farmer communities to implement continuity plans and ensure resilience in securing safe and nutritious food supply. It is important that such efforts and plans be in line with and support furthering the broader 2030 Agenda with targeted actions to accelerate progress to achieve various SDGs for better production, better nutrition, a better environment, and a better life, leaving no one behind.

GPSR Compliance
The European Union's (EU) General Product Safety Regulation (GPSR) is a set of rules that requires consumer products to be safe and our obligations to ensure this.

If you have any concerns about our products, you can contact us on

ProductSafety@springernature.com

In case Publisher is established outside the EU, the EU authorized representative is:

Springer Nature Customer Service Center GmbH
Europaplatz 3
69115 Heidelberg, Germany

www.ingramcontent.com/pod-product-compliance
Ingram Content Group UK Ltd.
Pitfield, Milton Keynes, MK11 3LW, UK
UKHW020934180926
475692UK00003B/32
9783031894398